站在巨人的肩上
Standing on Shoulders of Giants
TURING
图灵教育
iTuring.cn
U0310291

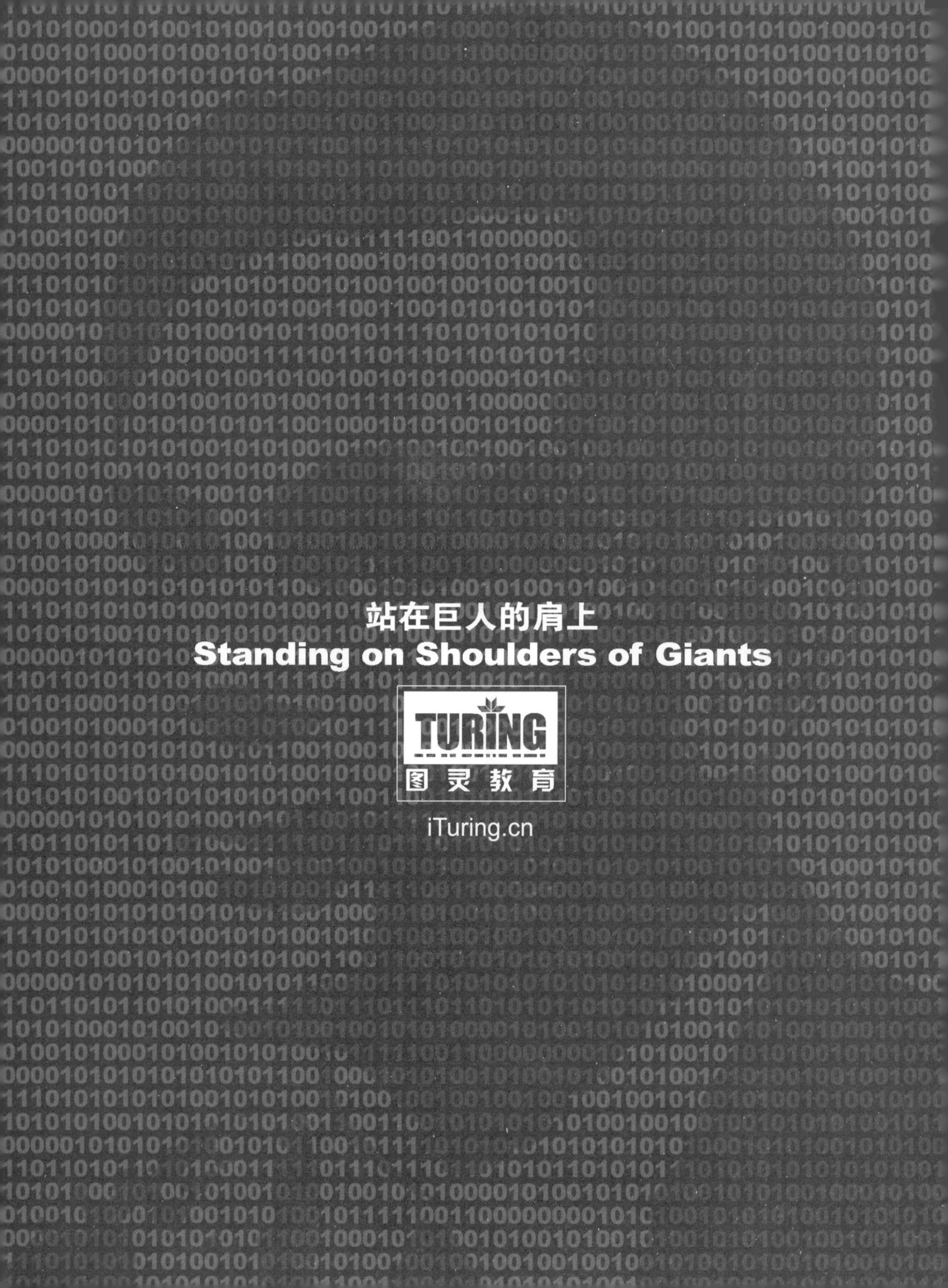
站在巨人的肩上
Standing on Shoulders of Giants
TURING
图灵教育
iTuring.cn

TURING 图灵程序设计丛书

JavaScript快速全栈开发

Rapid Prototyping with JS: Agile JavaScript Development

```
                  (
      )           )
      (     (     (
      )     )     )
###############
##################
###############   ##
 ##############   ##
 #################
   ##########
```

[美] Azat Mardanov◎著
胡波◎译

人民邮电出版社
北京

图书在版编目（CIP）数据

JavaScript快速全栈开发 / (美) 马尔达诺夫（Mardanov,A.）著 ; 胡波译. -- 北京 : 人民邮电出版社, 2015.1
（图灵程序设计丛书）
ISBN 978-7-115-37609-1

Ⅰ. ①J… Ⅱ. ①马… ②胡… Ⅲ. ①JAVA语言－程序设计 Ⅳ. ①TP312

中国版本图书馆CIP数据核字(2014)第273178号

内 容 提 要

本书涵盖 JavaScript 快速开发的多项前沿技术，是极其少见的前后端技术集大成之作。本书所涉技术包括 Node.js、MongoDB、Twitter Bootstrap、LESS、jQuery、Parse.com、Heroku 等，分三部分介绍如何用这些技术快速构建软件原型。第一部分是基础知识，让大家真正认识前后端及敏捷开发，并学会搭建本地及云环境。第二部分与第三部分分别介绍如何构建前端原型和后端原型。作者以前端组件开篇，通过为一个示例聊天应用 Chat 打造多个版本（Web/ 移动），将前端和后端结合在一起并给出应用部署方式。

本书适合进阶的初学者和中级 Web 及移动开发者阅读参考，特别适合熟悉 Ruby on Rails、PHP、Perl、Python 或者 Java 等语言的程序员。

◆ 著　　　　[美] Azat Mardanov
　译　　　　胡　波
　责任编辑　李松峰　毛倩倩
　执行编辑　姜力心
　责任印制　杨林杰

◆ 人民邮电出版社出版发行　　北京市丰台区成寿寺路11号
　邮编　100164　　电子邮件　315@ptpress.com.cn
　网址　http://www.ptpress.com.cn
　北京鑫正大印刷有限公司印刷

◆ 开本：800×1000　1/16
　印张：12.75
　字数：317千字　　　　2015年 1 月第 1 版
　印数：1 - 3 500册　　2015年 1 月北京第 1 次印刷

著作权合同登记号　图字：01-2014-3344号

定价：49.00元

读者服务热线：(010)51095186转600　印装质量热线：(010)81055316

反盗版热线：(010)81055315

广告经营许可证：京崇工商广字第 0021 号

版 权 声 明

读者反馈

“Azat的这本教程在我们为开发Sidepon.com网站构建的用户体验中起了至关重要的作用，同时它帮助我们优化了TheNextWeb.com，并成功盈利。”

——Kenson Goo，Sidepon.com

“阅读此书和其中的例子，让我得到了很多乐趣。相信它展示的例子也会帮助你发现大量的技术，这些技术是每个人都应该在自己项目里实践的。”

——Chema Balsas

本书已经被StartupMonthly①成功地用作培训②手册，这里是一些学员的感言。

“感谢大家，特别是Azat和Yuri。我很享受这个过程，现在有了极大的动力去努力了解这些技术。”

——Shelly Arora

“谢谢，和大家在一起的这个周末，我们使用Bootstrap和Parse做东西，简直太快捷，太神奇了！”

——Mariya Yao

“感谢Yuri以及所有人。这是一次很棒的聚会，非常有教育意义，对提升我的JavaScript技能有非常明显的作用。期待将来能与大家一起工作。”

——Sam Sur

---

① http://startupmonthly.org

② http://www.startupmonthly.org/rapid-prototyping-with-javascript-and-nodejs.html

网上资源

让我们在网上成为朋友，请通过以下方式关注我们。

- Twitter：@RPJSbook[①] 或者 @azat_co[②]
- Facebook：facebook.com/RapidPrototypingWithJS[③]
- 网址：rapidprototypingwithjs.com
- 博客：webapplog.com
- GitHub：github.com/azat-co/rpjs
- Storify：Rapid Prototyping with JS[④]

你还可以通过其他方式联系我们，如下。

- 电子邮件：hi@rpjs.co
- 谷歌用户组：rpjs@googlegroups.com或者https://groups.google.com/forum/#!forum/rpjs

① https://twitter.com/rpjsbook

② https://twitter.com/azat_co

③ https://www.facebook.com/RapidPrototypingWithJS

④ https://storify.com/azat_co/rapid-prototyping-with-js

致　谢

首先，感谢本书文字编辑David Moadel和技术编辑Alexander Vlasyuk。感谢我的学生们，他们可能来自Hack Reactor[1]、Marakana[2]、pariSOMA[3]和General Assembly[4]，在那里我用本书作为培训教材进行教学。

此外，我还要感谢StartupMonthly[5]团队，尤其是Yuri和Vadim。谢谢他们为我使用本书进行培训[6]及本书稿提供诸多宝贵的建设性意见。

最后，感谢我的设计师朋友们——Ben、Ivan 和 Natalie， 谢谢他们帮助我改进了封面设计。

① http://hackreactor.com
② http://marakana.com
③ http://pariSOMA.com
④ http://generalassemb.ly
⑤ http://startupmonthly.org
⑥ http://www.webapplog.com/training/

引 言

提要：使用快速原型以及写作本书的原因；回答本书是什么、不是什么以及阅读本书的前提条件；关于使用本书及相关示例的建议；解释书中的特定格式。

> “实践出真知。”
>
> ——史蒂夫 · 布兰克[1]

本书作为实践指南，介绍了如何使用最新的Web及移动技术快速构建软件原型。这些技术包括Node.js[2]、MongoDB[3]、Twitter Bootstrap[4]、LESS[5]、jQuery[6]、Parse.com[7]、Heroku[8]，等等。

为什么要撰写本书

其实本书是由失望激发的产物。作为一名具有多年工作经验的软件工程师，当我开始学习Node.js和Backbone.js时，发现从它们的官方文档入手相当困难，而且网上严重缺少快速入门指南和相应的示例。并且，你基本上不可能在同一个地方找到JS相关的高级技术的所有教程。

最好的学习方式就是实践，没错吧？因此我通过简单的小例子来实践，即快速入门指南，用来快速学习一些新技术。在完成一些基本的程序后，我需要一些参考文档和回顾。一开始我写这个指南只是自用，以加深对这些概念的理解，并且供以后参考。在StartupMonthly[9]我教了几次为期两天的集中课程，也是使用同样的理念，帮助有经验的开发者使用JavaScript进行敏捷开发。我们使用的手册得到了很多反馈，然后我们进行了大量更新。最终的成果就是你面前这本书了。

---

① http://steveblank.com/2010/03/11/teaching-entrepreneurship-%E2%80%93-by-getting-out-of-the-building/

② http://nodejs.org

③ http://mongodb.org

④ http://twitter.github.com/bootstrap

⑤ http://lesscss.org

⑥ http://jquery.com

⑦ http://parse.com

⑧ http://hreoku.com

⑨ http://startupmonthly.org

本书内容

正常情况下，读者一定会期待这里有一些快速入门指南、教程和建议（比如，Git工作流）。我们主要介绍如何编码，而非阐述理论知识，因此其中的理论会直接和实践部分相关，对于更好地理解相应技术（比如JSONP和跨域请求）以及用到的具体方法来说必不可少。

除了代码示例，本书几乎介绍了所有安装和部署步骤。

你将从前端组件开始，学习一个聊天（Web/移动）应用程序的例子。这个程序会有多个版本，而最终我们会把前端和后端结合到一起，然后将该程序发布到生产环境。这个聊天程序包含典型Web应用所有必要的组件，会帮你建立自己开发应用、应聘好工作或晋升，甚至是创业的信心。

目标读者

本书面向进阶的初学者和中级Web及移动开发者，即熟悉Ruby on Rails、PHP、Perl、Python或者Java等其他语言的专家。这类开发人员希望学习更多的JavaScript及Node.js相关技术来快速构建Web和移动程序原型，但可能没有时间去翻阅（大量或者哪怕些许）官方文章。我们并非想通过本书将读者成就为专家，而是希望帮助他们尽可能快地构建程序。

本书英文书名*Rapid Prototyping with JS: Agile JavaScript Development*直译为“用JS快速构建原型：JavaScript敏捷开发”，顾名思义，它就是要介绍如何用最快的速度以Web或者移动应用的形式构建出原型。这正是Lean Startup[①]里的思想，所以相对来说本书对于创业公司的创始人会更有意义，但大公司的员工同样会发现它的有用之处，特别是当他们想要掌握新技能，想要晋升或谋求更好的工作时。

这本书不是什么

这既不是一本全面介绍相关框架、库或者技术（或者某一特定技术）的书，也不是所有Web开发技术与技巧的参考书。本书中的例子很可能在网上有公开可用的类似源代码。

如果你不了解循环、条件判断语句、数组、散列、对象和函数等编程基础概念，请不要奢望在本书中了解它们。此外，理解书中的例子也将非常具有挑战性。

市面上已经有许多特别棒的书介绍了基本编程方法，本书最后就给出了一个此类书的列表，以方便大家查阅。再次提醒大家注意，本书的目的是讲述敏捷开发，而非重述编程理论和计算机科学知识。

---

① http://theleanstartup.com

先决条件

为更好地学习本书示例和相关内容，你需要：

- 关于编程的基本知识，比如对象、函数、数据结构（数组、散列）、循环（`for`、`while`）和条件判断（`if/else`、`switch`）;
- 基本的Web开发技能，包含但不仅限于HTML和CSS;
- 针对本书示例和一般Web开发我们强烈推荐Mac OS X 或者UNIX/Linux，不过你仍可以掌握基于Windows系统的敏捷开发技术;
- 能够访问网络;
- 5 ~ 20小时的时间 ;
- 信用卡或者借记卡，因为即使一些云服务的免费账户也需要它们。

示例

书中所有示例的源代码均可参见GitHub（http://github.com/azat-co/rpjs），你也可以下载ZIP文件[1]或者使用Git获取。更多关于怎样安装和使用Git的内容请参见本书后面的介绍。源代码文件、文件夹结构和部署文件理论上可以（或者稍加修改后）同时在本地和远程的云服务（PaaS，如Windows Azure和Heroku）上正常运行。

注意，GitHub上的代码和书中的代码在格式上可能稍有差别。另外，如果你发现任何bug，欢迎通过电子邮件（hi@rpjs.co）与我们联系。

格式说明

下面是源代码的样式:

```
var object = {};
object.name = "Bob";
```

终端命令与上面的格式差不多，只是以`$`开头:

```
$ git push origin heroku
$ cd /etc/
$ ls
```

行内术语采用楷体，命令名（如`mongod`）使用等宽字体。

① https://github.com/azat-co/rpjs/archive/master.zip

术语说明

本书中会使用可以互换的术语，尽管有时候在具体上下文中严格来说它们可能指代不同的东西。举几个例子：function = method = call（函数 = 方法 = 调用）、attribute = property = member = key（属性 = 特性 = 成员 = 键）、value = variable（值 = 变量）、object = hash = class（对象 = 散列 = 类）、list = array（列表 = 数组）、framework = library = module（框架 = 库 = 模块）。

目 录

第一部分 快速入门

第二部分 前端原型构建

Part 1

第一部分 快速入门

本部分内容

第1章

基础知识

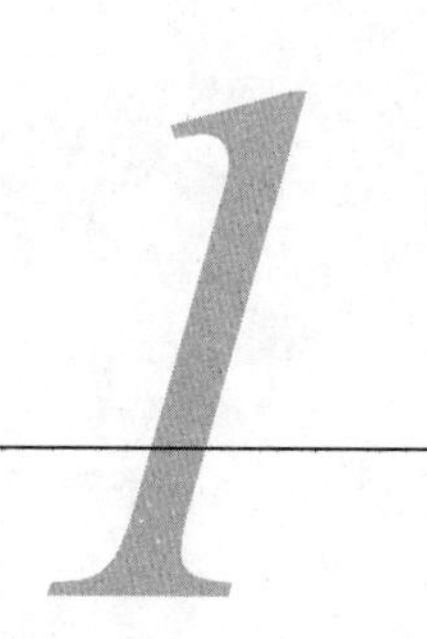

提要：综述 HTML、CSS 和 JavaScript 语法；简单介绍敏捷开发方法；云计算、Node.js 和 MongoDB 的好处；HTTP 请求/响应，以及 REST 式 API 的思想。

> 我认为每个人都应该学会为计算机编程，这会教你如何去思考。计算机科学是自由的艺术，是所有人都应该学习的。
>
> ——史蒂夫·乔布斯

1.1 定义前端

1.1.1 综述

Web 和移动应用开发过程一般包含以下步骤：

(1) 用户在浏览器（客户端）里输入或者点击一个链接；

(2) 浏览器向服务器发送 HTTP 请求；

(3) 服务器处理请求，如果查询字符串或者请求体里含有参数，服务器也会把这些参数信息考虑进去；

(4) 服务器更新、获取或者转换数据库里的数据；

(5) 服务器以 HTML、JSON 或者其他格式返回一个 HTTP 响应；

(6) 浏览器接收 HTTP 响应；

(7) 浏览器以 HTML 或者其他格式（比如 JPEG、XML 或者 JSON）把 HTTP 响应呈现给用户。

移动应用的行为动作与普通网站相同，只不过原生应用取代了浏览器。其他主要区别为：带宽带来的数据传输限制、更小的屏幕、更高效地使用本地存储。

这里有几种针对移动应用的开发方式，每种各有利弊：

- 用 Objective-C 和 Java 开发的原生 iOS 应用、Android 应用和 Blackberry 应用；
- 使用 JavaScript 开发，但是使用 Appcelerator[①] 或者类似的工具来编译为原生的 Objective-C

① http://www.appcelerator.com/

或 Java 应用；

- 通过响应式设计、Twitter Bootstrap[1]和 Foundation[2]等 CSS 框架、常规 CSS 或其他模板来适配小屏幕的移动网站；
- 使用 Sencha Touch[3]、Trigger.io[4]、JO[5]构建包含 HTML、CSS、JavaScript 的 HTML5 应用，然后使用 PhoneGap[6]包装成原生应用。

1.1.2 HTML

HTML 本质上不是编程语言，而是一组用来描述内容结构和格式的标记。HTML 标签由一对尖括号以及括号内的标签名组成。大多数情况下，内容会包含在一对开始标签和结束标签之间，结束标签的标签名前有一个斜杠（/）。

下面的例子里，每一行都是一个 HTML 元素：

```
<h2>Overview of HTML</h2>
<div>HTML is a ...</div>
<link rel="stylesheet" type="text/css" href="style.css" />
```

HTML 文档会有一个 `html` 标签，所有其他元素都是 `html` 标签的子标签：

```
<!DOCTYPE html>
<html lang="en">
  <head>
    <link rel="stylesheet" type="text/css" href="style.css" />
  </head>
  <body>
    <h2>Overview of HTML</h2>
    <p>HTML is a ...</p>
  </body>
</html>
```

HTML 有不同的版本，比如 DHTML、XHTML1.0、XHTML1.1、XHTML2、HTML4 和 HTML5。这篇文章对它们的区别做了很好的诠释：Misunderstanding Markup: XHTML 2/HTML 5 Comic Strip[7]。

所有的 HTML 元素都具备一些属性。重要的属性如下：class、id、style、data-name、onclick 以及其他事件属性。

---

① http://twitter.github.io/bootstrap/
② http://foundation.zurb.com/
③ http://www.sencha.com/products/touch/
④ http://trigger.io
⑤ http://joapp.com
⑥ http://phonegap.com
⑦ http://coding.smashingmagazine.com/2009/07/29/misunderstanding-markup-xhtml-2-comic-strip/

class

class 属性定义了一个类，以便于使用 CSS 给元素添加样式或者进行 DOM 操作，比如：

```
<p class="normal">...</p>
```

id

id 属性定义了元素的 ID，作用有点像 class，但是必须是唯一的，比如：

```
<div id="footer">...</div>
```

style

style 属性定义了一个元素的内联 CSS， 比如：

```
<font style="font-size:20px">...</font>
```

title

title 属性为元素指定了一些额外信息，在大多数浏览器里这些信息均是以小提示条的形式呈现的。比如：

```
<a title="Up-vote the answer">...</a>
```

data-name

data-name 属性可以用来在 DOM 中存储一些元数据。比如：

```
<tr data-token="fa10a70c-21ca-4e73-aaf5-d889c7263a0e">...</tr>
```

onclick

onclick 属性意味着在点击事件发生时，内联的 JavaScript 代码将运行，比如：

```
<input type="button" onclick="validateForm();">...</a>
```

onmouseover

和 onclick 属性类似，但它由鼠标悬停事件触发，比如：

```
<a onmouseover="javascript: this.setAttribute('css','color:red')">...</a>
```

其他与内联 JavaScript 代码相关的 HTML 属性如下。

- onfocus：当浏览器的焦点聚集在某个元素上时触发。
- onblur：当浏览器的焦点离开一个元素时触发。
- onkeydown：用户按下键盘上的键时触发。
- ondblclick：用户双击鼠标时触发。

- onmousedown：用户按下鼠标时触发。
- onmouseup：用户释放鼠标时触发。
- onmouseout：用户将鼠标移开元素区域时触发。
- oncontextmenu：用户点击鼠标右键时触发。

完整的事件列表和浏览器兼容性表格请参见“Event compatibility tables”[1]。

我们将会广泛使用 Twitter Bootstrap 里的类，而由于内联 CSS 和 JavaScript 不是好方案，我们会尽量避免内联。不管怎样，了解一下 JavaScript 事件名总不会错，因为在 jQuery、Backbone.js 和纯 JavaScript 中我们常常会用到它们。把内联的属性转换为 JS 事件，只需要把 on 前缀去掉就行了，比如 onclick 属性就是指 click 事件。

你还可以从以下三个网站看到更多相关资料：Catching a mouse click[2]、Wikipedia[3]和 w3schools[4]。

1.1.3 CSS

CSS 是一种控制内容呈现和格式的方式。HTML 文档可以通过一个 link 标签引入外部的 CSS 文件（如前面的例子所示），也可以直接通过 style 标签内联 CSS 代码，比如：

```
<style>
  body {
    padding-top: 60px; /* 上内边距为 60 像素 */
  }
</style>
```

每个 HTML 元素都可以拥有 id 和/或 class 属性：

```
<div id="main" class="large">
  Lorem ipsum dolor sit amet,
  Duis sit amet neque eu.
</div>
```

在 CSS 里，我们可以通过元素的 id、class、标签名，以及它与父级标签的关系或者元素属性值来定位元素。

下面的规则把所有的段落（p 标签）的颜色变成了灰色（#999999）：

```
p {
  color:#999999;
}
```

① http://www.quirksmode.org/dom/events/index.html
② https://developer.mozilla.org/en-US/docs/JavaScript/Getting_Started#Example:_Catching_a_mouse_click
③ http://en.wikipedia.org/wiki/HTML
④ http://www.w3schools.com/html/html_intro.asp

下面的规则通过 id main 定位了一个 div，并且设置了它的内边距：

```
div#main {
  padding-bottom:2em;
  padding-top:3em;
}
```

下面的规则把所有拥有类 large 的元素字体大小设置为 14pt：

```
.large {
  font-size:14pt;
}
```

div 是 body 元素的直接子元素；现在要隐藏下标签 div：

```
body > div {
  display:none;
}
```

设置 name 属性为 email 的 input 元素的宽度为 150px：

```
input[name="email"] {
  width:150px;
}
```

更多信息可以参考 Wikipedia[①]和 w3schools[②]。

CSS3 是 CSS 的一个升级版，它可以直接用 CSS 实现圆角、边框和渐变，而 CSS 如果想运用这些功能，只能依靠 PNG/GIF 的帮助或者使用其他一些小技巧才能实现。

更多信息可以参考 CSS3.info[③]、w3school[④]，以及对比 CSS3 和 CSS 的文章“CSS3 vs. CSS: A Speed Benchmark”[⑤]。

1.1.4 JavaScript

JavaScript 是 1995 年在 Netscape 以 LiveScript 开始的。它和 Java 的关系就像雷锋与雷峰塔的关系一样，两者风马牛不相及。现在，JavaScript 在 Web 开发的前后端都有使用，在桌面应用开发中也占有一席之地。

把 JS 代码放进 HTML 文档的 script 标签里是使用 JavaScript 的最简单方式：

---

① http://en.wikipedia.org/wiki/Cascading_Style_Sheets

② http://www.w3schools.com/css/

③ http://css3.info

④ http://www.w3schools.com/css/

⑤ http://coding.smashingmagazine.com/2011/04/21/css3-vs-css-a-speed-benchmark/

```
<script type="text/javascript" language="javascript">
  alert("Hello world!");
  //简单的提示对话框
</script>
```

我们建议你不要把 HTML 与 JS 代码混在一起，为了把它们分离开，可以把代码移到一个外部文件中，然后再通过设置 `script` 标签的源属性 `src="filename.js"`来引入外部的 js 文件，这里我们用 `src="app.js"`：

```
<script src="js/app.js" type="text/javascript" language="javascript">
</script>
```

注意

结束标签`</script>`必须存在，即使像我们这样引入外部文件后它是一个没有内容的标签。经过多年的发展，JavaScript 已经成为了浏览器脚本里的绝对主导者，而 `Type` 和 `language` 已经不是必须存在的了。

其他运行 JavaScript 代码的方式：

- 内联的代码（之前已经讲述过）；
- 使用 Webkit 的浏览器开发者工具和 FireBug 控制台；
- Node.js 的交互命令行。

JavaScript 的一个优点就是它是弱类型的。弱类型是相对于强类型[①]语言（如 C 和 Java）而言的，这使 JavaScript 成为了一个更好的原型语言。下面是 JavaScript 对象/类（本身没有类；对象继承自对象）里一些主要的类型：

数值原始值

数值，比如：

```
var num = 1;
```

数值对象

数值[②]对象和它的方法，比如：

```
var numObj = new Number("123"); //数值对象
var num = numObj.valueOf(); //数值原始值
var numStr = numObj.toString(); //字符串表示
```

字符串原始值

包含在单引号或者双引号之间的字符序列，比如：

---

① http://en.wikipedia.org/wiki/Strong_typing

② https://developer.mozilla.org/en-US/docs/JavaScript/Reference/Global_Objects/Number

```
var str = "some string";
var newStr = "abcde".substr(1,2);
```

为了方便，JS 自动给字符串原始值包装上字符串对象方法，但是它们并不完全相同[①]的。

字符串对象

字符串对象有很多非常有用的方法，比如 length、match，来看个例子：

```
var strObj = new String("abcde");//字符串对象
var str = strObj.valueOf(); //字符串原始值
strObj.match(/ab/);
str.match(/ab/); //两种调用都可行
```

正则表达式对象

正则表达式对象是特殊的字符模式，以方便搜索、替换以及测试字符串：

```
var pattern = /[A-Z]+/;
str.match(/ab/);
```

特殊类型

当你有疑问的时候，可以使用 typeof 对象来看看它的类别。下面是 JS 里的一些特殊类型：

- NaN
- null
- undefined
- function

全局方法

你可以在代码的任意地方调用这些方法，因为它们是全局方法：

- decodeURI
- decodeURIComponent
- encodeURI
- encodeURIComponent
- eval
- isFinite
- isNaN
- parseFloat
- parseInt

① https://developer.mozilla.org/en-US/docs/JavaScript/Reference/Global_Objects/String#Distinction_between_string_primitives_and_String_objects

- uneval
- Infinity
- Intl

JSON

JSON 库帮助我们序列化和解析 JavaScript 对象，比如：

```
var obj = JSON.parse('{a:1, b:"hi"}');
var stringObj = JSON.stringify({a:1,b:"hi"});
```

数组对象

数组[①]是从 0 开始索引的列表。例如，创建一个数组：

```
var arr = new Array();
var arr = ["apple", "orange", 'kiwi"];
```

数组有一些非常好用的方法，比如 indexOf、slice、join。你要确保自己对它们非常熟悉，因为如果能够正确使用，它们将帮你节省很多时间。

数据对象

```
var obj = {name: "Gala", url:"img/gala100x100.jpg",price:129}
```

或

```
var obj = new Object();
```

下面是一些继承模式。

布尔原始值和对象

就像字符串和数值，布尔值[②]既可以是原始值，也可以是对象。

```
var bool1 = true;
var bool2 = false;
var boolObj = new Boolean(false);
```

日期对象

日期[③]对象帮助我们处理日期和时间，比如：

```
var timestamp = Date.now(); // 1368407802561
var d = new Date(); //Sun May 12 2013 18:17:11 GMT-0700 (PDT)
```

---

① https://developer.mozilla.org/en-US/docs/JavaScript/Reference/Global_Objects/Array

② https://developer.mozilla.org/en-US/docs/JavaScript/Reference/Global_Objects/Boolean

③ https://developer.mozilla.org/en-US/docs/JavaScript/Reference/Global_Objects/Date

数学对象

数学常量和一些方法[①]，比如：

```
var x = Math.floor(3.4890);
var ran = Math.round(Math.random()*100);
```

浏览器对象

用于访问浏览器及其一些属性，比如 URL，来看个例子：

```
window.location.href = 'http://rapidprototypingwithjs.com';
console.log("test");
```

DOM 对象

```
document.write("Hello World");
var table = document.createElement('table');
var main = document.getElementById('main');
```

警告

JavaScript 支持的数字大小上限为 53 位数，如果你需要处理比这个大的数字，那么最好使用一些大数字库。

你可以在 Mozilla Developer Network[②]和 w3school[③]网站上找到完整的 JavaScript 和 DOM 对象参考。

如果你还想了解 JS 的一些资源，比如 ECMA 规范，请查看这个列表：JavaScript Language Resources[④]。本书撰写之时，最新的 JavaScript 规范是 ECMA-262 Edition 5.1: PDF[⑤]和 HTML[⑥]。

函数式和原型语言是 JS 的另一个重要特性。一般的函数声明语法是这样的：

```
function Sum(a, b) {
  var sum = a + b;
  return sum;
}
console.log(Sum(1, 2));
```

由于函数式编程[⑦]特性，函数在 JS 里是一等公民[⑧]。函数可以用作变量和对象，比如，一个函数可以作为另一个函数的参数传递：

---

① https://developer.mozilla.org/en-US/docs/JavaScript/Reference/Global_Objects/Math

② https://developer.mozilla.org/en-US/docs/JavaScript/Reference

③ http://www.w3schools.com/jsref/default.asp

④ https://developer.mozilla.org/en-US/docs/JavaScript/Language_Resources

⑤ http://www.ecma-international.org/publications/files/ECMA-ST/Ecma-262.pdf

⑥ http://www.ecma-international.org/ecma-262/5.1/

⑦ http://en.wikipedia.org/wiki/Functional_programming

⑧ http://en.wikipedia.org/wiki/First-class_function

```
var f = function(str1) {
  return function(str2) {
  return str1 + ' ' + str2;
  };
};
var a = f('hello');
var b = f('goodbye');
console.log((a('Catty'));
console.log((b('Doggy'));
```

了解几种在 JS 里中的对象继承方式也很有用：

- 类式继承[①]
- 伪类继承[②]
- 函数式继承

如果你想进一步了解继承方式，请查看“Inheritance Patterns in JavaScript”[③]和“Inheritance revisited”[④]。

更多关于浏览器里的 JS 的资料可以从网站 Mozilla Developer Network[⑤]、Wikipedia[⑥] 和 w3schools[⑦]中找到。

1.2 敏捷开发概述

由于像瀑布流这样的传统软件开发方法在不确定性很高的情况下（即解决办法未知[⑧]）难以很好地工作，因此敏捷软件开发方法应运而生，一点一滴发展起来。敏捷方法包括 Scrum/Sprint、测试驱动开发、持续集成、并行开发和其他实践方法，很多都是借鉴极限编程。

1.2.1 Scrum

对于管理，敏捷使用 Scrum 方法。如果想了解更多关于 Scrum 的知识，请阅读：

- Scrum Guide（PDF）[⑨]；
- Scrum.org[⑩]；

① http://www.crockford.com/javascript/inheritance.html
② http://javascript.info/tutorial/pseudo-classical-pattern
③ http://bolinfest.com/javascript/inheritance.php
④ https://developer.mozilla.org/en-US/docs/JavaScript/Guide/Inheritance_Revisited
⑤ https://developer.mozilla.org/en-US/docs/JavaScript/Reference
⑥ http://en.wikipedia.org/wiki/JavaScript
⑦ http://www.w3schools.com/js/default.asp
⑧ http://www.startuplessonslearned.com/2009/03/combining-agile-development-with.html
⑨ http://www.scrum.org/storage/scrumguides/Scrum_Guide.pdf
⑩ http://www.scrum.org/

❑ 维基百科上关于Scrum软件开发的文章[①]。

Scrum由一个个短周期组成，每个周期叫sprint。一个sprint通常持续一到两周。典型的sprint是在sprint计划会议上开始和结束的，这些会议同时会把新任务分配给团队成员。新任务不能添加到正在进行的sprint里，它们只能在sprint计划会议上添加。

每天的scrum会议是Scrum方法体系的重要组成部分，Scrum正是因此得名。每一个scrum通常是在走廊里进行的5 ~ 15分钟的会议。在scrum会上，每一个团队成员要回答下面三个问题：

(1) 自己昨天做了什么；

(2) 今天准备做什么；

(3) 是否需要从其他团队成员那里得到些什么。

与瀑布流开发相比，敏捷开发更加灵活，特别是在高度变化的环境（创业初期）中。

敏捷开发的好处：在没法提前规划时，或者需要收集用户反馈作为主要决策因素时，敏捷开发非常高效。

1.2.2 测试驱动开发

测试驱动开发，也叫TDD，包含下面的步骤：

(1) 使用断言（`true`或`false`）为新功能、新任务或者增强写下失败的自动测试用例；

(2) 写出能够成功通过测试的代码；

(3) 如果需要，重构代码，添加功能，同时保证测试用例通过；

(4) 重复上面的步骤直到所有任务都完成。

测试可以分为功能测试和单元测试。单元测试通过模拟依赖测试系统单元、函数和方法，功能测试（也叫集成测试）是包含依赖的情况下测试一系列功能。

测试驱动开发的好处：

❑ 更少的错误和缺陷；

❑ 更有效的代码；

❑ 确保代码可以正常工作，并且不会影响已有功能。

1.2.3 持续部署和集成

持续部署，也叫CD，是一系列技术的组合，可以快速把开发好的新功能、错误修复和增强功能呈现给用户。CD包含自动测试和自动部署。持续部署可以减少我们的手工劳动，尽可能地缩短收集反馈的时间。一般来说，开发者越快地从用户那里收集反馈，就能越快地增强产品，这对于竞争来讲非常有利。很多创业公司一天就部署很多次，相比之下，有些成熟的大公司一次发

① http://en.wikipedia.org/wiki/Scrum_(development)

布周期就要半年甚至一年的时间。

持续部署的好处：减少反馈时间和手工劳动。

关于持续部署和持续集成的区别，请参考这篇文章：Continuous Delivery vs. Continuous Deployment vs. Continuous Integration - Wait huh?[①]。

下面是流行的持续集成解决方案。

- Jenkins[②]：开源的可扩展持续集成服务器。
- CircleCI[③]：更好更快的代码。
- Travis CI[④]：一个针对开源社区的持续集成托管服务。

1.2.4 结对编程

结对编程是两个开发者在同一个环境里一起工作的技术。其中一个开发者为“驾驶员”，另一个为“观察员”。驾驶员主要负责写代码，观察员围观并且提供建议。一段时间后两者互换角色。驾驶员的角色更多关注当下的任务，观察员则需要有大局观，发现错误并且改进算法。

结对编程的优势：

- 两人可以一起产出更简洁、更高效的代码，同时降低引发错误和缺陷的几率；
- 此外还有一个附加的好处，那就是程序员们在一起工作时彼此可以交流和传递知识。但是，程序员间也可能会发生冲突，而且这并不少见。

1.3 后端定义

1.3.1 Node.js

Node.js 是事件驱动异步 I/O 开源程序，可以用来创建可伸缩且高性能 Web 服务器。Node.js 由谷歌的 V8 JavaScript 引擎[⑤]组成，由云服务公司 Joyent[⑥]维护。

Node.js 项目的本意及其使用就像 Twisted[⑦]之于 Python，或者 EventMachine[⑧]之于 Ruby 那样。

---

① http://blog.assembla.com/assemblablog/tabid/12618/bid/92411/Continuous-Delivery-vs-Continuous-Deployment-vs-Continuous-Integration-Wait-huh.aspx

② http://jenkins-ci.org/

③ https://circleci.com/

④ https://travis-ci.org/

⑤ http://en.wikipedia.org/wiki/V8_(JavaScript_engine)

⑥ http://joyent.com

⑦ http://twistedmatrix.com/trac/

⑧ http://rubyeventmachine.com/

用 JavaScript 来实现 Node 是继 Ruby 和 C++语言之后的第三次尝试。

Node.js 本身并非 Ruby on Rails 那样的框架，它和 PHP+Apache 的相似度更高。在第 6 章，我们将详尽介绍 Node.js 的框架。

使用 Node.js 有如下优势。

- JavaScript 语言作为 Web 和移动开发的业界标准，开发人员熟悉他的可能性更高。
- 前后端用同一个编程语言进行开发，可以加快写代码的过程。开发者的思维不需要在不同的语法间切换（上下文切换），而且他们可以更快地学习各种方法和类。
- 使用 Node.js 可以更快速地搭建原型，做市场开发，提前获取用户。与使用 PHP 或者 MySQL 等不太敏捷的技术的公司相比，这是很重要的优势。
- Node.js 可以通过 Web-sockets 支持实时应用程序。

如果你想了解更多信息，请阅读 Wikipedia[1]、Nodejs.org[2]、ReadWrite[3]及 O'Reilly[4]上的文章。

如果你想了解我们写作本书时 Node.js 的情况，请参考 Isaac Z. Schlueter 于 2013 年制作的幻灯片"State of the Node"[5]。

1.3.2 NoSQL和MongoDB

MongoDB 来自于 huMONGOus，是可以用于大数据的高性能非关系型数据库。由于传统的关系型数据库管理系统（RDBMS）无法应对大数据的挑战，NoSQL 概念应运而生。

MongoDB 有如下优势。

- 可伸缩性：由于其分布性特性，可以在多个服务器和数据中心放置冗余数据。
- 高性能：MongoDB 在存储和检索方面非常高效，部分原因是数据库中元素与集合之间关系上的弱处理。
- 灵活性：键-值对存储非常适合原型开发，因为它使用户不需要关心表关系，不需要修复数据模型，不需要关注复杂的数据迁移。

1.3.3 云计算

云计算由下列服务组成：

---

① http://en.wikipedia.org/wiki/Nodejs

② http://nodejs.org/about/

③ http://readwrite.com/2011/01/25/wait-whats-nodejs-good-for-aga

④ http://radar.oreilly.com/2011/07/what-is-node.html

⑤ http://j.mp/2013-state-of-the-node

- 基础设施即服务（IaaS），比如 Rackspace、Amazon Web Services；
- 平台即服务（PaaS），比如 Heroku、Windows Azure；
- 后端即服务（BaaS，最新、最酷的一种），比如 Parse.com、Firebase；
- 软件即服务（SaaS），比如 Google Apps、Saleforce.com。

云应用平台具备以下功能：

- 可伸缩性，例如在几分钟内产生新的实例；
- 部署简单，即向 Heroku 部署，只需要`$ git push`；
- 现收现付制，用户根据需要添加和删除内存及磁盘空间；
- 可扩展，方便安装和设置数据库、应用服务器、程序包等；
- 安全和技术支持。

PaaS 和 BaaS 非常适合构建原型，一般用来创建 MVP（Minimal Viable Product，最小化且可行的产品），供那些尚处于创业初始阶段的团队使用。

下面是一些为人熟知的 PaaS 提供商：

- Heroku①
- Windows Azure②
- Nodejitsu③
- Nodester④

1.3.4 HTTP请求和响应

每一个 HTTP 请求和响应均由下面这些组件构成。

(1) 头部（header）：关于编码、主体的长度、来源、内容类型等的信息。
(2) 主体（body）：内容，一般是参数或者数据，通常传递给服务器，或者返回给客户端。

另外，HTTP 请求包含以下几方面内容。

- 方法：一些请求方法，比如常见的 `GET`、`POST`、`PUT`、`DELETE`。
- URL：主机、端口和路径，比如 https://graph.facebook.com/498424660219540。
- 查询字符串：URL 里问号之后的所有字符，比如?q=rpjs&page=20。

---

① http://heroku.com
② http://windowsazure.com
③ http://nodejitsu.com/
④ http://nodester.com

1.3.5　REST式API

由于在分布式系统中每个事务都需要包含足够多关于客户端状态的信息，REST（REpresentational State Transfer）式 API 因此流行起来。从某种意义上来说，这个标准也是无状态的，因为客户端的状态并不会保存在服务器上，这样才使得每一个请求可以分发到不同的服务系统上进行处理。

REST 式 API 的特征：

- 有更好的可伸缩性，因为它支持把不同的组件部署到不同的服务器上；
- 替代 SOAP（Simple Object Access Protocol，简单对象访问协议），因为它简单的动词和名词组合；
- 充分利用 HTTP 方法，例如 `GET`、`POST`、`DELETE`、`PUT`、`OPTIONS`，等等。

下面是一个简单的 REST 式 API 的例子，它包含了对消息的创建、读取、更新、删除（CRUD）功能：

方　法	URL	含　义
GET	/messages.json	以JSON格式返回消息列表
PUT	/messages.json	更新或者替换所有的消息，以JSON格式返回状态或者错误信息
POST	/messages.json	创建一个新消息，以JSON格式返回它的ID
GET	/messages/{id}.json	以JSON格式返回ID为{id}的某个消息
PUT	/messages/{id}.json	替换或者更新ID为{id}的消息，如果不存在就创建
DELETE	/messages/{id}.json	删除ID为{id}的消息，以JSON格式返回状态或者错误信息

REST 不是一种协议，而是一种比诸如 SOAP 这样的协议更灵活的架构。因此，当我们需要获得一些格式方面的支持时，它的URL可以形如`/messages/list.html`或者`/messages/list.xml`。

`PUT` 和 `DELETE` 是幂等方法[1]，这意味着如果服务器接收到两个或者更多个类似的请求，返回结果是一样的。

`GET` 是幂零的，`POST` 是非幂等的，它们可能影响状态并且导致副作用。

更多关于 REST API 的内容，敬请参阅 Wikipedia[2]和“A Brief Introduction to REST article”[3]一文。

---

① http://en.wikipedia.org/wiki/Hypertext_Transfer_Protocol#Idempotent_methods_and_web_applications

② http://en.wikipedia.org/wiki/Representational_state_transfer

③ http://www.infoq.com/articles/rest-introduction

第2章

设　置

提要：给出关于工具集的一些建议；逐步安装本地组件；为使用云服务做准备。

> 我最有成效的一天就是扔掉了1000行代码。
>
> ——肯·汤普森[1]

2.1　本地环境搭建

2.1.1　开发目录

如果你没有特定的用来做Web开发的目录，建议在Documents（中文系统里一般显示为文档）目录里创建名为Development的目录，它的路径将会是Documents/Development。使用示例代码，在刚创建的目录里再创建一个 rpjs 目录，它的路径将会是 Documents/Development/rpjs，可以在Mac OS X上使用Finder或者在OS X/Linux系统上使用下面的命令：

```
$ cd ~/Documents
$ mkdir Development
$ cd Development
$ mkdir rpjs
```

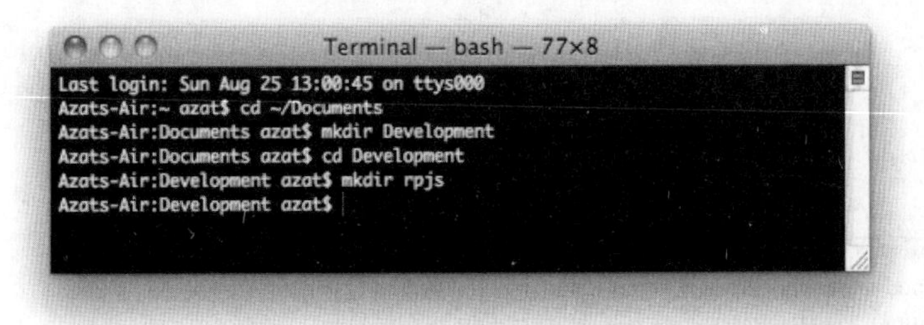

初始开发环境设置

① http://en.wikipedia.org/wiki/Ken_Thompson

提示

在 Mac OS 上如果想通过终端在当前目录里打开 Finder 应用，只需输入并执行 `$ open .`命令。

查看文件和目录，使用下面的命令：

```
$ ls
```

把隐藏的文件和目录（例如.git）也显示出来，使用下面的命令：

```
$ ls -lah
```

除了`$ ls`，我们还可以使用`$ ls -alt`。`-lah` 和`-alt` 参数的不同之处在于，前者是按文件字母排序，后者按文件时间排序。

注意

你可以使用 Tab 键来自动完成文件和目录名。

接下来，你可以复制示例到 rpjs 目录并创建应用。

注意

另一件有用的事情是设置 Finder 的“新建位于文件夹位置的终端”选项。打开“系统设置”（可以通过“Command + 空格”使用 Spotlight 搜索），找到“键盘”并点击，打开“键盘快捷键”，点击“服务”，勾选“新建位于文件夹位置的终端标签”和“新建位于文件夹位置的终端窗口”设置快捷键即可。

2.1.2　浏览器

我们建议你下载最新的 Webkit[①]或者 Gecko[②]浏览器：Chrome[③]、Safari[④]或者 Firefox[⑤]。Chrome 和 Safari 已经内置了开发者工具，Firefox 需要安装插件 Firebug[⑥]。

① http://en.wikipedia.org/wiki/WebKit

② http://en.wikipedia.org/wiki/Gecko_(layout_engine)

③ http://www.google.com/chrome

④ http://www.apple.com/safari/

⑤ http://www.mozilla.org/en-US/firefox/new/

⑥ http://getfirebug.com/

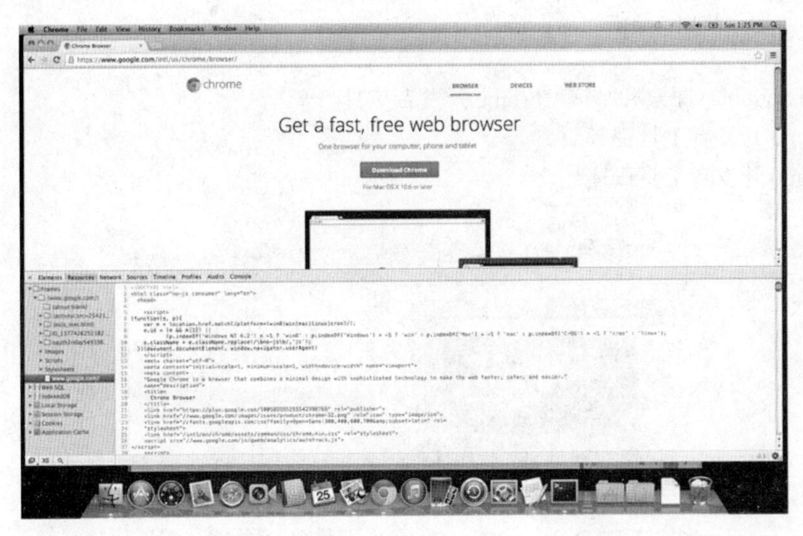

Chrome 开发者工具

Firebug 和开发者工具可以帮助开发者完成诸如下面的事情：

- 调试 JavaScript；
- 修改 HTML 和 DOM 元素；
- 实时修改 CSS；
- 监控 HTTP 请求和响应；
- 运行性能分析，查看堆转储；
- 查看已经加载的资源，如图片、CSS 和 JS 文件。

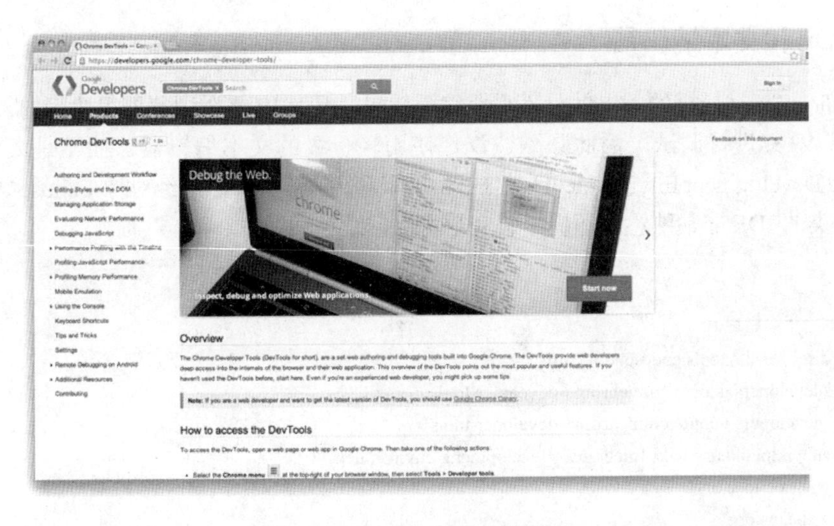

关于 Web Deb 工具的谷歌教程

非常棒的 Chrome 开发者工具教程有：

- Code School，探索和掌握 Chrome 开发者工具[①]；
- Chrome 开发者工具视频[②]；
- Chrome 开发者工具概述[③]。

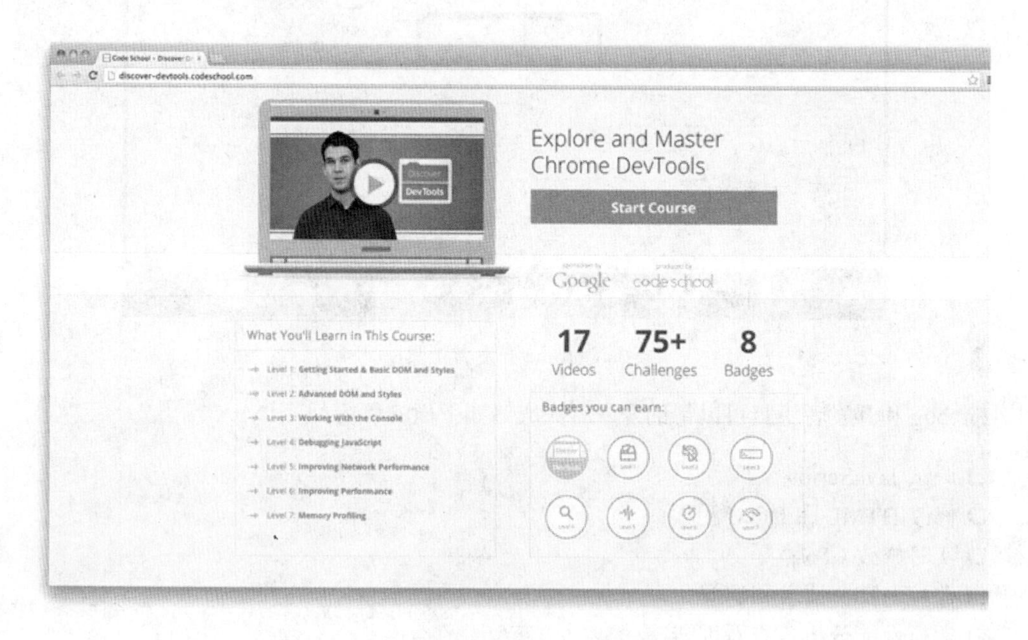

掌握 Chrome 开发者工具

2.1.3 IDE和文本编辑器

使用 JavaScript 最棒的一点就是不需要编译代码。因为 JS 是在浏览器里执行的，所以在浏览器里就可以对其进行调试。因此强烈建议你使用轻量级的文本编辑器，而不是大而全的 IDE（Integrated Development Environment[④]，集成开发环境），但是如果你已经对像 Eclipse[⑤]、NetBeans[⑥]、Aptana[⑦]之类的 IDE 非常熟悉和适应了，完全可以继续使用。

---

① http://discover-devtools.codeschool.com/
② https://developers.google.com/chrome-developer-tools/docs/videos
③ https://developers.google.com/chrome-developer-tools/
④ http://en.wikipedia.org/wiki/Integrated_development_environment
⑤ http://www.eclipse.org/
⑥ http://netbeans.org/
⑦ http://aptana.com/

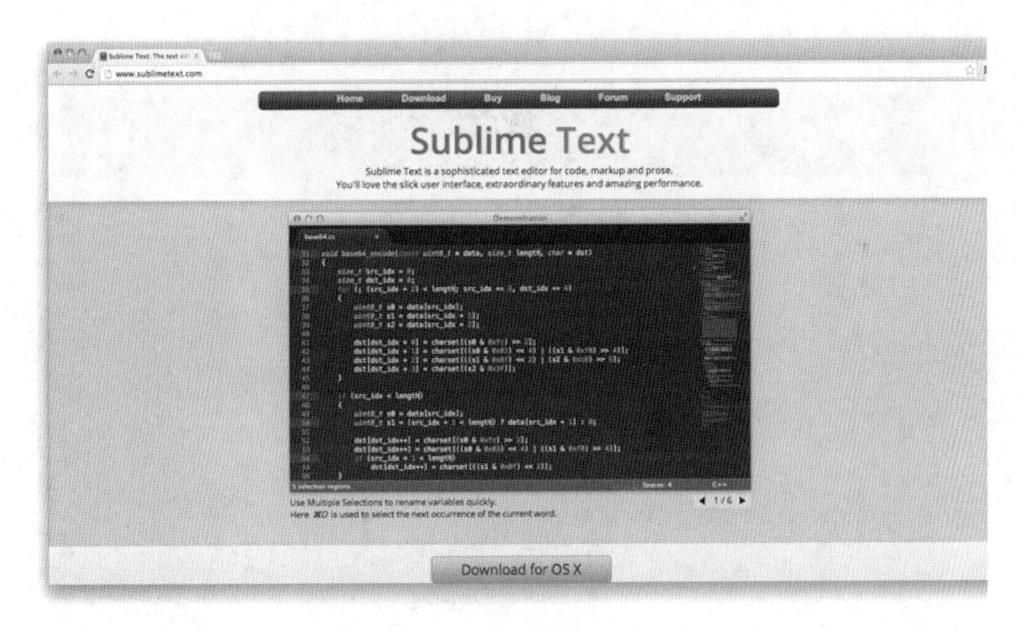

Sublime Text 代码编辑器首页

下面是一些常用于 Web 开发的文本编辑器和 IDE。

- TextMate[1]：只有 Mac OS X 版本，1.5 版本可免费试用 30 天，号称 The Missing Editor for Mac OS X。
- Sublime Text[2]：Mac OS X 和 Windows 版本都有，是比 TextMate 更好的编辑器，无限期试用。
- Coda[3]：一体化的编辑器，带有 FTP 浏览器和预览，支持使用 iPad 进行开发。
- Aptana Studio[4]：全面的 IDE，包含内建的终端和其他很多工具。
- Notepad ++[5]：Windows 特有的开源轻量级文本编辑器，支持多种语言。
- WebStorm IDE[6]：功能丰富的 IDE，可以做 Node.js 调试；由 JetBrains 开发，称为“最智能的 JavaScript IDE”。

---

① http://macromates.com/
② http://www.sublimetext.com/
③ http://panic.com/coda/
④ http://aptana.com/
⑤ http://notepad-plus-plus.org/
⑥ http://www.jetbrains.com/webstorm/

WebStorm IDE 首页

2.1.4 版本控制系统

版本控制系统[①]在只有一个开发者的情况下也应该使用。很多云服务，例如 Heroku，部署时也需要 Git。同时强烈建议你使用 Git 和 Git 终端命令，而不是带有图形用户界面的 Git 可视化客户端/应用，例如 GitX[②]、Gitbox[③]、GitHub for Mac[④]。

Subversion 是一个非分布式版本控制系统。GitSvnComparison[⑤]这篇文章对比了 Git 和 Subversion。

下面是在机器上安装和设置 Git 的步骤。

(1) 从 http://git-scm.com/downloads 下载最新版的适用于你的操作系统的 Git。

(2) 利用下载的`*.dmg`包安装 Git，即运行`*.pkg`文件，并按照步骤操作。

(3) 在 OS X 上通过“Command + 空格”打开 Spotlight 搜索（见下面的截图），查找并打开终端。在 Windows 上可以使用 PuTTY[⑥]或者 Cygwin[⑦]。

---

① http://en.wikipedia.org/wiki/Revision_control

② http://gitx.frim.nl/

③ http://www.gitboxapp.com/

④ http://mac.github.com/

⑤ https://git.wiki.kernel.org/index.php/GitSvnComparison

⑥ http://www.chiark.greenend.org.uk/~sgtatham/putty/

⑦ http://www.cygwin.com/

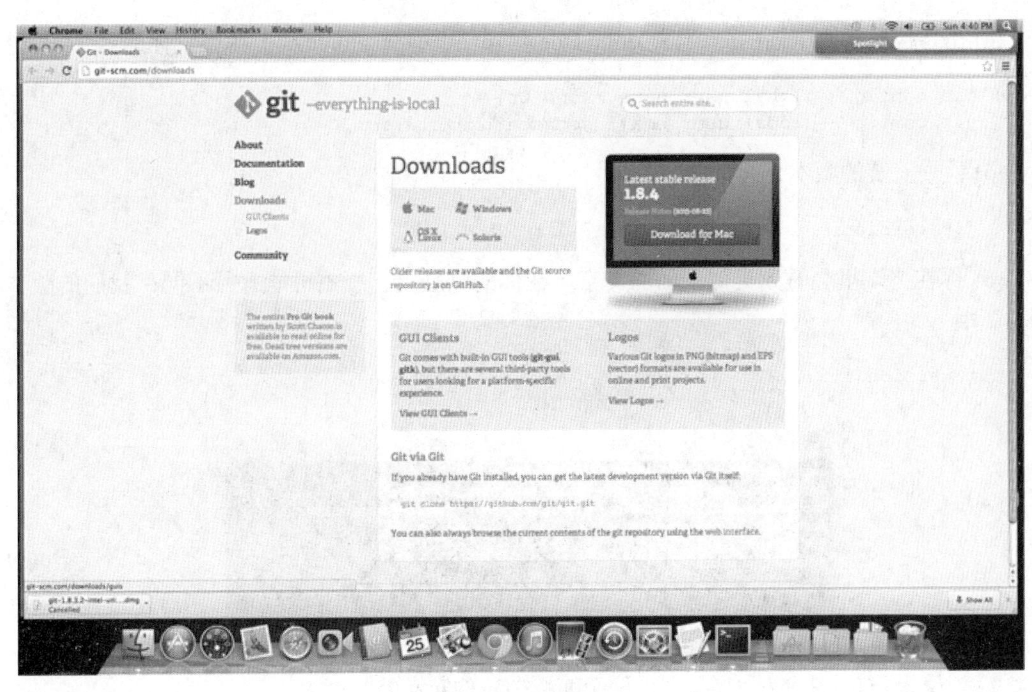

下载最新版 Git

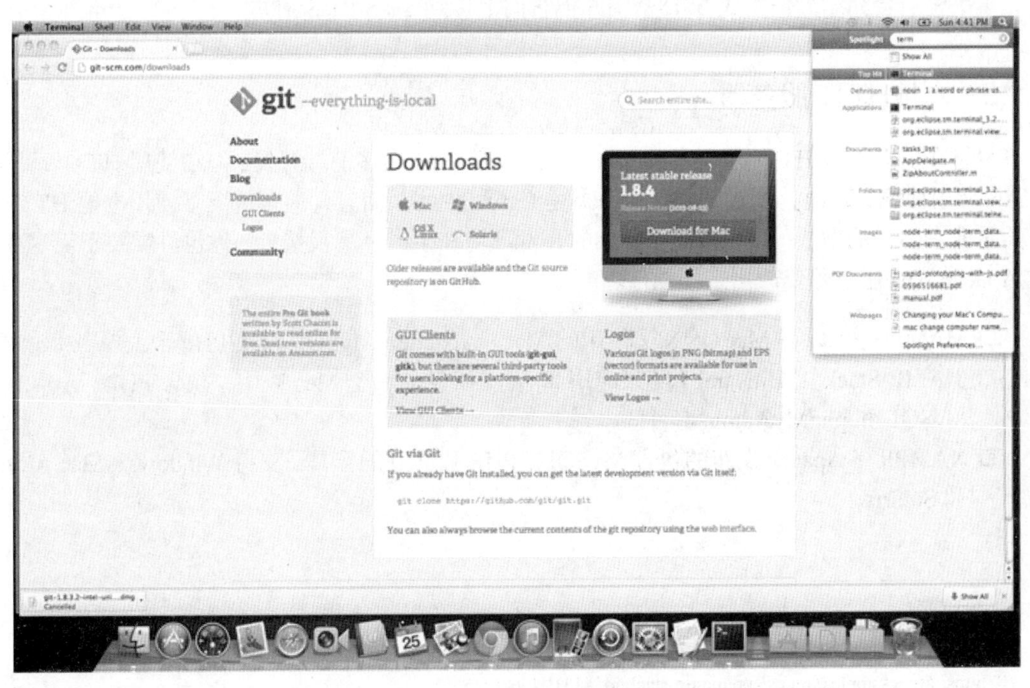

使用 Spotlight 查找并且执行应用

(4) 在终端里输入下面的命令，把 John Doe 和 johndoe@example.com 替换成你的名字和邮箱：

```
$ git config --global user.name "Jonh Doe"
$ git config --global user.email johndoe@example.com
```

(5) 检查安装的版本，运行命令：

```
$ git version
```

(6) 你在终端里会看到类似下面的文字（在你的电脑上，版本可能有所不同，我们的版本是 1.8.3.2）：

```
git version 1.8.3.2
```

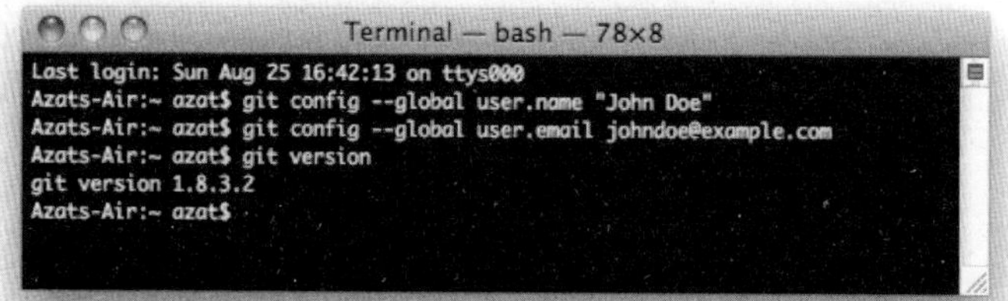

```
Terminal — bash — 78×8
Last login: Sun Aug 25 16:42:13 on ttys000
Azats-Air:~ azat$ git config --global user.name "John Doe"
Azats-Air:~ azat$ git config --global user.email johndoe@example.com
Azats-Air:~ azat$ git version
git version 1.8.3.2
Azats-Air:~ azat$
```

设置和测试 Git 的安装

本书稍后会介绍如何生成 SSH 密钥并把它们上传到 SaaS/PaaS 网站。

2.1.5 本地HTTP服务器

尽管在没有本地 HTTP 服务器的情况下依然可以完成大多前端开发，但在使用 HTTP 请求和 AJAX 调用加载文件时，本地 HTTP 服务器是必不可少的。而且，通常情况下使用本地 HTTP 服务器是非常不错的做法。这种情况下，你的开发环境和生产环境很接近。你也许需要下面这些通过 Apache Web 服务器修改出来的软件。

- MAMP[①]：可以在 Mac OS X 上使用的 Mac、Apache、MySQL 以及 PHP 个人 Web 服务器。
- MAMP Stack[②]：BitNami（苹果应用商店[③]）研发的一个 Mac 应用，包含 PHP、Apache、MySQL 和 phpMyAdmin 套件。
- XAMPP[④]：Apache 发布的包含 MySQL、PHP 和 Perl 的工具，支持 Windows、Mac、Linux 和 Solaris。

① http://www.mamp.info/en/index.html
② http://bitnami.com/stack/mamp
③ https://itunes.apple.com/es/app/mamp-stack/id571310406?l=en
④ http://www.apachefriends.org/en/xampp.html

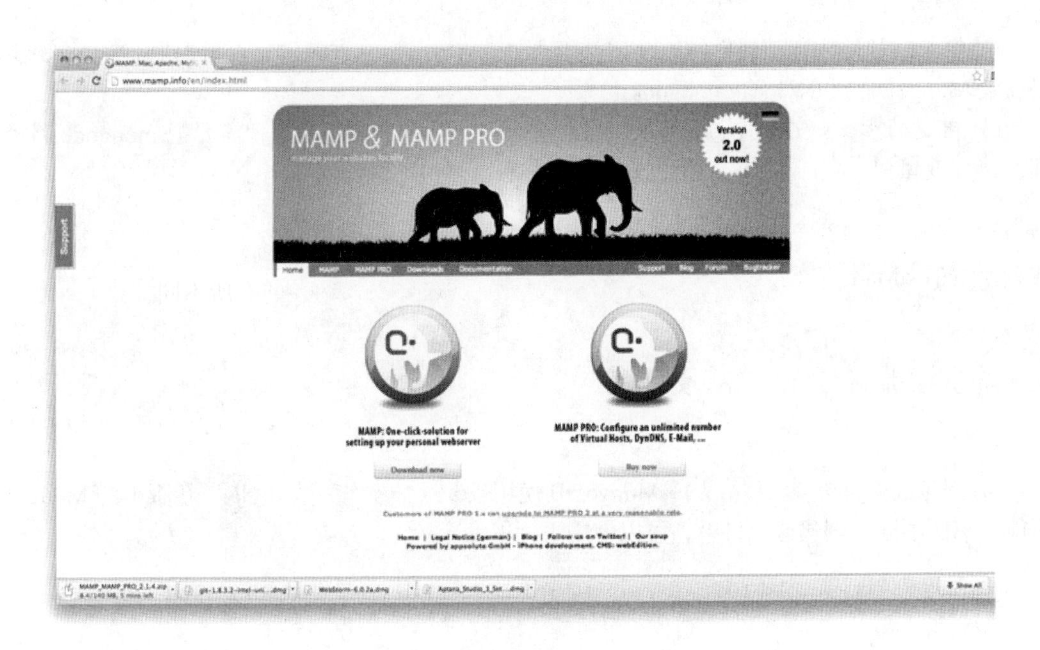

Mac 版 MAMP 主页

MAMP、MAMP Stack 和 XAMPP 都拥有直观的图形用户界面（GUI），可以通过它们来修改配置和主机文件设置。

注意

Node.js 和其他很多后端技术一样，拥有自己的用于开发环境的服务器。

2.1.6 数据库：MongoDB

下面的步骤更适合基于 Mac OS X 或者 Linux 的系统，但经过修改也可用于 Windows 系统，即修改第(3)步里的`$PATH`变量。接下来，我们将详细描述如何使用官方安装包来安装 MongoDB，因为我们发现这种方法更健壮，引入的冲突也更少。当然，还有很多其他在 Mac 以及其他系统[①]上安装的方式[②]，例如使用 Brew。

(1) 可以从 http://www.mongodb.org/downloads 下载 MongoDB。最新的苹果笔记本，像 MacBook Air，可以选择 OS X 64-bit 版本。老版本的 Mac 可以查看链接：http://dl.mongodb.org/dl/osx/i386。

提示

可以通过输入命令`$ uname -p`查看你的电脑处理器是什么架构体系。

---

① http://docs.mongodb.org/manual/installation/

② http://docs.mongodb.org/manual/tutorial/install-mongodb-on-os-x/

(2) 将下载的文件解压到你的 Web 开发目录中(~/Documents/Development 或其他地方)，也可以安装 MongoDB 到/usr/local/mongodb 目录。

(3) 可选的步骤：如果你想在系统的任何位置访问 MongoDB 命令，需要把 mongodb 路径加到`$PATH`变量里。在 Mac OS X 中打开 `paths` 路径需使用命令：

```
sudo vi /etc/paths
```

或者使用 TextMate 打开：

```
mate /etc/paths
```

把下面的命令加到`/etc/paths`文件里：

```
/usr/local/mongodb/bin
```

(4) 创建数据目录。默认情况下，MongoDB 使用`/data/db`作为数据目录。新版本的 MongoDB 可能会有所不同。创建这个目录，可以使用下面的命令：

```
$ sudo mkdir -p /data/db
$ sudo chown `id -u` /data/db
```

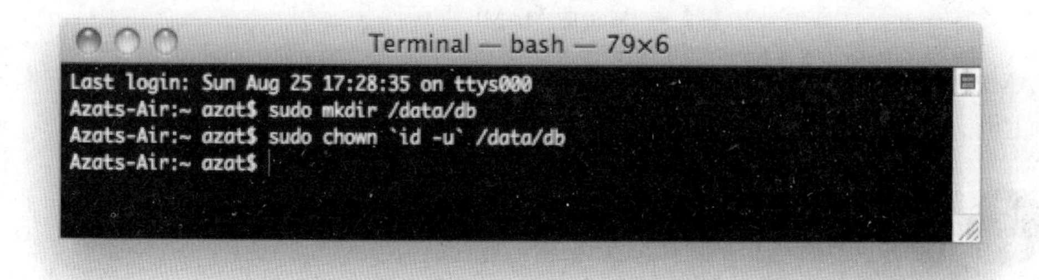

初始化设置 MongoDB：创建数据目录

如果你不喜欢使用`/data/db`而是希望指定一个目录，可以通过 mongod 的`--dbpath`参数来指定（MongoDB 主要服务）。

(5) 进入解压 MongoDB 的目录。这个目录中会有一个 bin 目录，我们在终端里使用下面的命令：

```
$ ./bin/mongod
```

```
Terminal — mongod — 122×30
Last login: Sun Aug 25 17:29:16 on ttys000
Azats-Air:~ azat$ mongod
mongod --help for help and startup options
Sun Aug 25 17:31:00
Sun Aug 25 17:31:00 warning: 32-bit servers don't have journaling enabled by default. Please use --journal if you want dur
ability.
Sun Aug 25 17:31:00
Sun Aug 25 17:31:00 [initandlisten] MongoDB starting : pid=738 port=27017 dbpath=/data/db/ 32-bit host=Azats-Air.local
Sun Aug 25 17:31:00 [initandlisten]
Sun Aug 25 17:31:00 [initandlisten] ** NOTE: This is a development version (2.3.0) of MongoDB.
Sun Aug 25 17:31:00 [initandlisten] **       Not recommended for production.
Sun Aug 25 17:31:00 [initandlisten]
Sun Aug 25 17:31:00 [initandlisten] ** NOTE: This is a 32 bit MongoDB binary.
Sun Aug 25 17:31:00 [initandlisten] **       32 bit builds are limited to less than 2GB of data (or less with --journal).
Sun Aug 25 17:31:00 [initandlisten] **       Note that journaling defaults to off for 32 bit and is currently off.
Sun Aug 25 17:31:00 [initandlisten] **       See http://www.mongodb.org/display/DOCS/32+bit
Sun Aug 25 17:31:00 [initandlisten]
Sun Aug 25 17:31:00 [initandlisten] ** WARNING: soft rlimits too low. Number of files is 256, should be at least 1000
Sun Aug 25 17:31:00 [initandlisten]
Sun Aug 25 17:31:00 [initandlisten] db version v2.3.0, pdfile version 4.5
Sun Aug 25 17:31:00 [initandlisten] git version: 86d6c3b316da2fffc1001e665442ba679b51fd26
Sun Aug 25 17:31:00 [initandlisten] build info: Darwin bs-osx-106-i386-1.local 10.8.0 Darwin Kernel Version 10.8.0: Tue Ju
n 7 16:33:36 PDT 2011; root:xnu-1504.15.3~1/RELEASE_I386 i386 BOOST_LIB_VERSION=1_49
Sun Aug 25 17:31:00 [initandlisten] options: {}
Sun Aug 25 17:31:00 [initandlisten] Unable to check for journal files due to: boost::filesystem::directory_iterator::const
ruct: No such file or directory: "/data/db/journal"
Sun Aug 25 17:31:00 [websvr] admin web console waiting for connections on port 28017
Sun Aug 25 17:31:00 [initandlisten] waiting for connections on port 27017
```

启动 MongoDB 服务器

(6) 如果你看到类似下面的输出：

```
MongoDB starting: pid = 7218 port=27017...
```

这意味着 MongoDB 数据库服务器已经启动了。默认情况下，它监听 http://localhost:27017。如果你用浏览器打开 http://localhost:28017，那么可以看到版本号、日志及其他有用的信息。在这个例子里 MongoDB 服务器使用了两个不同的端口（27017 和 28017）：一个是主（本机）端口，用来和别的应用交换数据，另一个是基于 Web 的图形用户界面，用来监控 Web 界面/统计数据。在 Node.js 代码里只需要使用 27017。

注意

修改$PATH 变量后不要忘记重新启动终端窗口。

现在，更进一步，需要测试我们是否可以访问 MongoDB 控制台/shell，它将作为这个服务器的一个客户端。为此，我们需要保持运行服务器的终端窗口处于打开状态。

(1) 在相同的目录里另开一个终端窗口，并且执行：

```
$ ./bin/mongo
```

你应该会看到类似这样的输出："MongoDB shell version 2.0.6 ..."。

(2) 然后输入并且执行：

```
> db.test.save({a:1})
> db.test.find()
```

如果看到添加的记录能被再次查询出来，就证明一切正常：

```
Terminal — mongo — 79×14
Last login: Sun Aug 25 17:30:33 on ttys000
Azats-Air:~ azat$ mongo
MongoDB shell version: 2.3.0
connecting to: test
Welcome to the MongoDB shell.
For interactive help, type "help".
For more comprehensive documentation, see
        http://docs.mongodb.org/
Questions? Try the support group
        http://groups.google.com/group/mongodb-user
> db.test.save( { a: 1} )
> db.test.find()
{ "_id" : ObjectId("521a169d6421d0d4d6f3190f"), "a" : 1 }
>
```

运行 MongoDB 客户端并且保存数据

find 和 save 命令做的事情和你想让它们做的事情一致。

更详细的介绍可以在 MongoDB.net 查阅：Install MongoDB on OS X[①]。Windows 用户可以阅读这篇文章：Installing MongoDB[②]。

注意

MAMP 和 XAMPP 包含的是 MySQL，它是开源的传统 SQL 数据库；phpMyAdmin 是 MySQL 数据库的 Web 管理应用。

注意

在 Mac OS X 和大部分 Unix 系统中，可以使用 control + c 关闭进程。如果使用 control + z，进程会休眠（或退出终端窗口）。这种情况下，数据文件会有一个锁文件，你必须通过 kill 命令或者活动监视器来关闭进程，然后手工删除数据文件夹里的锁文件。在 Mac 终端里 command + . 和 control + c 的效果是一样的。

2.1.7　其他组件

1. Node.js 安装

从 http://nodejs.org/#download 下载 Node.js，参见下面的截图。安装非常简单：下载压缩文件，

① http://docs.mongodb.org/manual/tutorial/install-mongodb-on-os-x/

② http://www.tuanleaded.com/blog/2011/10/installing-mongodb/

运行`*.pkg`安装器。检查 Node.js 的安装，可以使用下面的命令：

```
$ node -v
```

正常情况下会看到类似下面的信息（你的版本有可能不太一样）：

```
v0.8.1
```

Node.js 包含了 NPM（Node 包管理器[1]），我们会经常使用 NPM 来安装 Node.js 模块。

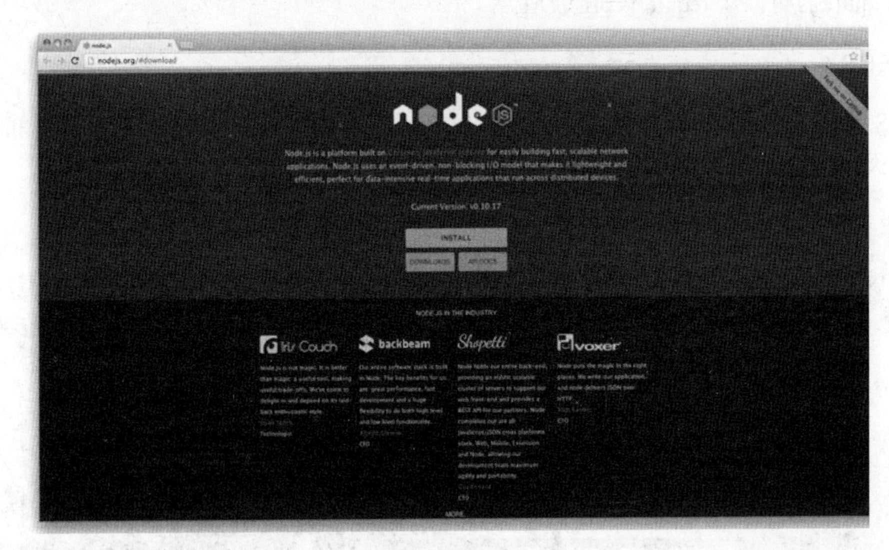

Node.js 主页

2. JS 库

前端 JS 库一般从它们的网站上下载并解压到开发目录里（如/Documents/Development），以备将来使用。它们经常会有压缩过的用于生产环境的版本（更多相关信息请参见 4.6 节）和有大量详细注释的用于开发环境的版本。

另一个办法是直接使用开放的 CDN 服务，例如 Google Hosted Libraries[2]、CDNJS[3]、Microsoft Ajax Content Delivery Network[4]，等等。使用这个办法后，一些用户使用应用时速度会快一些，但是如果没有网络，应用是完全不可用的。

下面是一些常用的前端 JS 库及它们的网址。

---

① https://npmjs.org

② https://developers.google.com/speed/libraries/devguide

③ http://cdnjs.com/

④ http://www.asp.net/ajaxlibrary/cdn.ashx

- LESS 是一个前端解释器，从 lesscss.org 获取。你可以解压它到你的开发目录（~/Documents/Development）或者其他地方。
- Twitter Bootstrap 是一个 CSS/LESS 框架，可以从 twitter.github.com/bootstrap 获取。
- jQuery 可以从 jquery.com 获取。
- Backbone.js 可以从 backbonejs.org 获取。
- Underscore.js 可以从 underscorejs.org 获取。
- Require.js 可以从 requirejs.org 获取。

3. LESS App

LESS App 是 Mac OS X 上的应用，它可以实时编译 LESS 文件到 CSS，可以从 incident57.com/less 获取。

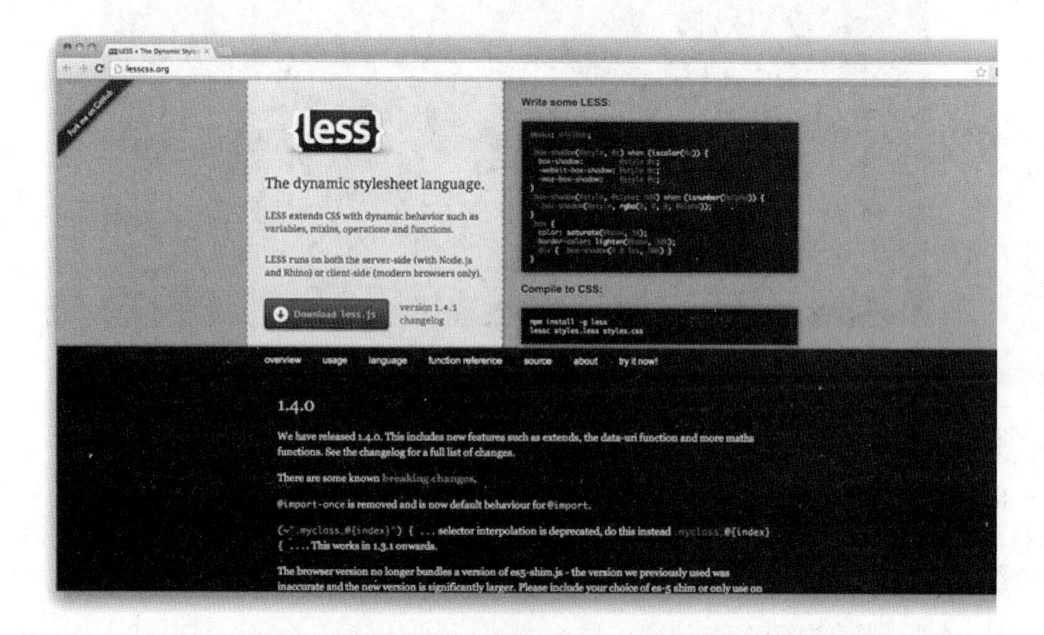

LESS App 首页

2.2　云端环境搭建

2.2.1　SSH密钥

使用 SSH 密钥可以安全连接到服务器，而不需要每次都输入用户名和密码。对于 GitHub 上的仓库，后一种登录方式使用的是 HTTPS URL，比如 https://github.com/azat-co/rpjs.git，前一种方式使用的是 SSH URL，比如 git@github.com:azat-co/rpjs.git。

在 Mac OS X 或 Unix 操作系统上产生 SSH 密钥可以按照下面的步骤操作。

(1) 检查已经存在的 SSH 密钥：

```
$ cd ~/.ssh
$ ls -lah
```

(2) 如果你看到有文件名字类似 id_rsa（参考截图），删除它们，或者使用下面的命令把它们移到备份目录里：

```
$ mkdir key_backup
$ cp id_rsa* key_backup
$ rm id_rsa*
```

(3) 使用 ssh-keygen 命令生成一对新的 SSH 密钥，假设在~/.ssh 目录：

```
$ ssh-keygen -t rsa -C "your_email@youremail.com"
```

(4)回答它提出的问题，最好保持默认的名字 id_rsa，然后复制 id_rsa.pub 里的内容：

```
$ pbcopy < ~/.ssh/id_rsa.pub
```

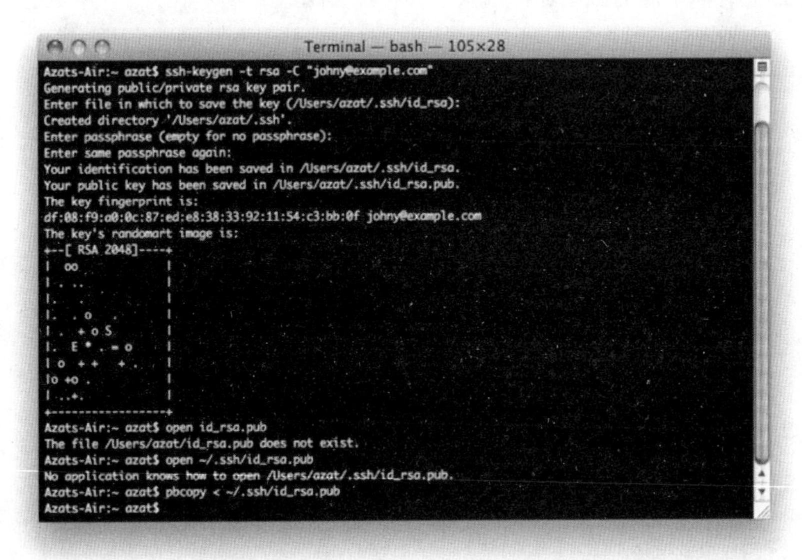

生成 SSH RSA 密钥并且复制公钥到剪切板

(5) 或者使用系统默认的编辑器打开 id_rsa.pub 文件：

```
$ open id_rsa.pub
```

(6) 也可以使用 TextMate：

```
$ mate id_rsa.pub
```

2.2.2 GitHub

(1) 复制公钥后，打开并登录 github.com，打开账户设置，选择“SSH key”然后添加新的 SSH 密钥。添加一个名字，比如你的电脑名称，然后粘贴复制的公钥。

(2) 检查是否可以通过 SSH 连接到 GitHub，输入并执行下面的命令：

```
$ ssh -T git@github.com
```

如果看到类似下面的内容：

```
Hi your-GitHub-username! You've successfully authenticated, but GitHub does not
provide shell access.
```

说明已经设置好了。

(3) 第一次连接到 GitHub 时会收到一个“authenticityofhost...can't be established”提醒。对于这个提示不用困惑，像下面的截图那样输入“yes”就行了。

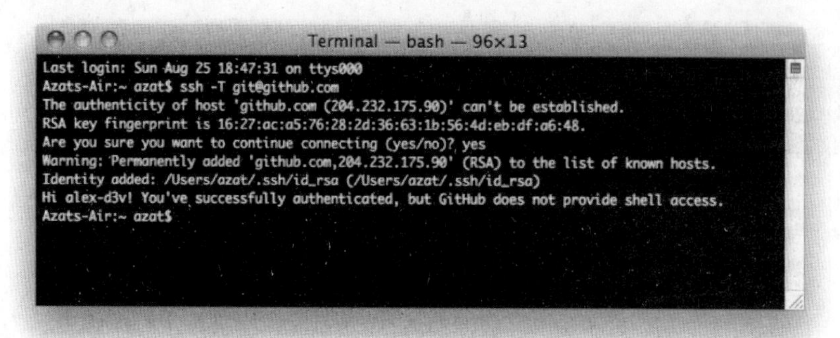

首次测试 SSH 连接到 GitHub

如果展示的不是类似的信息，请重复上一节内容里的步骤(3)~步骤(4)，然后在 GitHub 上更新产生的*.pub 文件里的内容。

警告

保护你的 id_rsa 隐私，不要把它共享给任何人。

更多说明可以在 GitHub 上找到：Generation SSH Keys（生成 SSH 密钥）[1]。

① https://help.github.com/articles/generating-ssh-keys

Windows 用户可以在[PuTTY]里找到生成 SSH 密钥的功能。

2.2.3 Windows Azure

下面是设置 Windows Azure 账户的步骤。

(1) 使用 Windows Azure 网站和虚拟机预览需要注册。目前，它们有 90 天的免费试用期：https://www.windowsazure.com/en-us。

(2) 启用 Git 部署，并且创建用户名和密码，然后上传 SSH 公钥到 Windows Azure。

(3) 安装 Node.js SDK，它可以从 https://www.windowsazure.com/en-us/develop/nodejs/获取。

(4)检查安装类型：

```
$ azure -v
```

正常情况下你会看到下面的输出：

```
Windows Azure: Microsoft's Cloud Platform... Tool Version 0.6.0
```

(5) 登录 Windows Azure 网站，网址是 https://windows.azure.com。

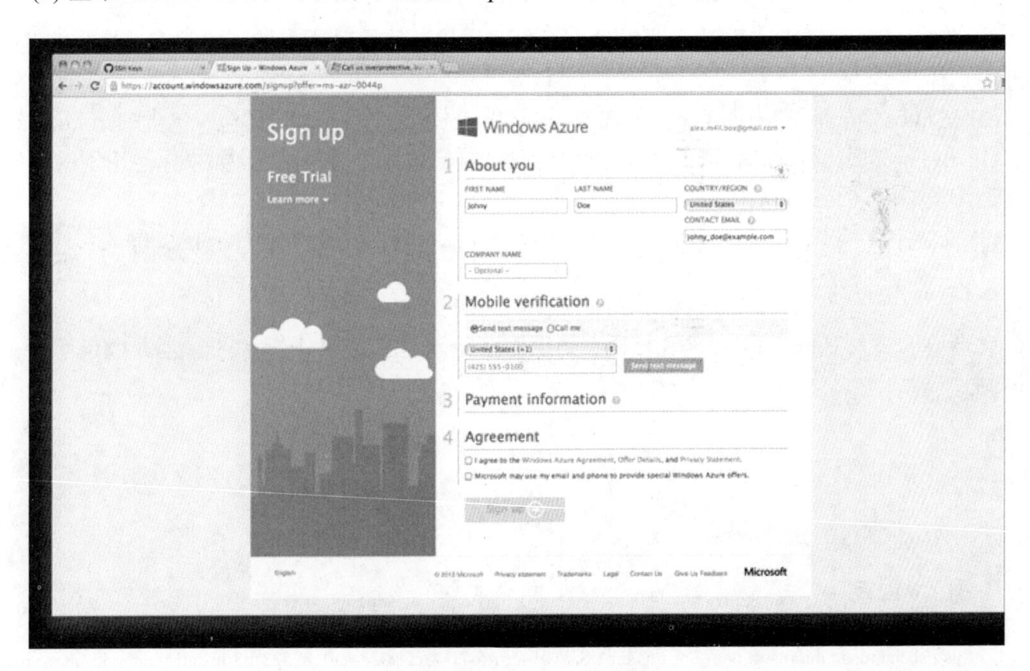

在 Windows Azure 上注册

(6) 选择“New”，然后选择“Web Site”和“Quick Create”。输入你期望使用的网址，然后点击“OK”。

(7) 到新创建的网站的控制面板，选择“Set up Git publishing”。然后，你需要输入用户名和密码。这个操作会应用到你的所有网站上，它意味着所有你创建的网站都不需要设定凭据。点击“OK”。

(8) 接下来，它会要求你填入一个 Git URL，比如：

```
https://azatazure@azat.scm.azurewebsites.net/azat.git
```

接下来是如何部署，本书后面的章节会有讲解。

(9) 高级用户选项：按照下面的这个教程创建一个虚拟机，并且在它上面安装 MongoDB：*Install MongoDB on a virtual machine running CentOS Linux in Windows Azure*[①]。

2.2.4 Heroku

Heroku（http://www.heroku.com）是多语言敏捷应用部署的平台。它工作方式类似于 Windows Azure，也可以通过 Git 部署应用。不需要为 MongoDB 安装虚拟机，因为它有 MongoHQ 扩展[②]。

我们可以按照下面的步骤设置 Heroku。

(1) 在 http://heroku.com 注册。当前他们有免费用户，使用时所有的选项都应选最小的（0），数据库选择共享。

(2) 在 https://toolbelt.heroku.com 下载 Heroku Toolbelt。它是一个工具包，包含 Heroku 拥有的库、Git 和 Foreman[③]等。老版本的 Mac 用户可以直接使用这个客户端[④]。如果你使用别的操作系统，请浏览 Hero Client GitHub[⑤]。

(3) 安装完成后，应该可以使用 heroku 命令。检查一下，并且登录 Heroku，输入：

```
$ heroku login
```

它会要求输入 Heroku 用户名和密码，如果你已经创建了 SSH 密钥，它会自动上传到 Heroku 的网站。

① https://social.msdn.microsoft.com/Forums/en-US/61014ef6-fd71-4aef-a02d-aa0182523fb3/install-mongodb-on-a-virtual-machine-running-centos-linux-in-windows-azure?forum=WAVirtualMachinesforWindows

② https://addons.heroku.com/mongohq

③ https://github.com/ddollar/foreman

④ http://assets.heroku.com/heroku-client/heroku-client.tgz

⑤ https://github.com/heroku/heroku

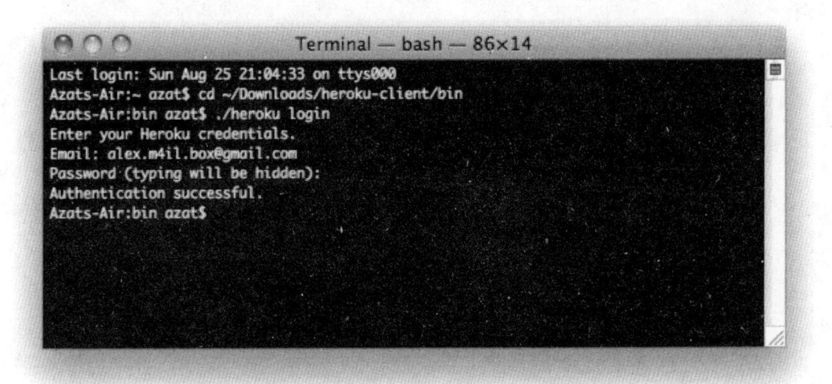

成功使用了`$ heroku login`命令的响应

(4) 如果一切正常，在特定的目录创建一个 Heroku 应用，执行下面的命令：

```
$ heroku create
```

更多步骤操作请参见 Heroku: Quickstart[①]和 Node.js[②]。

2.2.5 Cloud9

Cloud9 是一个在浏览器里运行的 IDE，绑定 GitHub 或者 BitBucket 账户后，它可以用于浏览、编辑你的仓库，并且部署到 Windows Azure 或者其他的服务。它不需要安装，所有的东西都在浏览器里工作，和 Google Docs 很类似。

① https://devcenter.heroku.com/articles/quickstart

② https://devcenter.heroku.com/articles/nodejs

Part 2

第二部分

前端原型构建

本部分内容

第3章

jQuery 和 Parse.com

提要：简述 jQuery 的主要函数，Twitter Bootstrap 支架（scaffolding），以及主要的 LESS 组件；JSON、AJAX 和 CORS 的定义；在 Twitter REST API 示例上演示 JSONP 调用；详解如何使用 jQuery 和 Parse.com 构建纯 Chat 前端应用；逐步介绍如何部署到 Heroku 和 Windows Azure。

> 世上有两种软件设计方法，一种是设计得尽量简单，以至于明显没有什么缺陷，另外一种是使它尽量得复杂，以至于其缺陷不那么明显。第一种方式更难。
>
> ——托尼·霍尔[1]

3.1 定义

3.1.1 JSON

这是从 json.org[2]摘抄的关于 JSON 的定义：

> JSON（JavaScript Object Notation）是一种轻量级的数据交换格式，易于人类读写，也易于机器解析和生成。它的产生是基于 JavaScript 编程语言（1999 年 12 月，标准 ECMA-262，第 3 版[3]）的一个子集。JSON 采用完全独立于语言的文本格式，但是也采用了为编程人员所熟知的 C 语言家族（包括 C、C++、C#、Java、JavaScript、Perl、Python 等）的约定。这些特性使 JSON 成为理想的数据交换语言。

JSON 已经成为 Web 和移动应用不同组件间、同第三方服务间数据交换的标准格式。在应用内部，JSON 也是广泛使用的格式，它可以用来存储设置信息、本地化、翻译文件或其他数据。

JSON 对象一般是这样的：

```
{
  "a": "value of a",
```

① http://en.wikipedia.org/wiki/Charles_Antony_Richard_Hoare

② http://www.json.org/

③ http://www.ecma-international.org/publications/files/ECMA-ST/Ecma-262.pdf

```
  "b": "value of b"
}
```

这里定义了一个包含键–值对的对象。键和值分别位于“:”号两边。在计算机科学术语中，JSON 等同于散列表，即一个用键值标识的列表或关联数组，具体叫什么取决特定的语言。JSON 和 JS 对象字面上最大的区别在于 JSON 更严格，必须使用“双引号”包裹键和字符串值。这两种类型都可以使用 `JSON.stringify()`序列化成一个字符串，使用 `JSON.parse()`反序列化一个有效的 JSON 对象字符串。

大家应该记得一个对象的成员可以是数组、数字或者其他对象，比如：

```
{
  "posts": [{
    "title": "Get your mind in shape!",
    "votes": 9,
    "comments": ["nice!", "good link"]
    }, {
    "title": "Yet another post",
    "votes": 0,
    "comments": []
    }
  ],
  "totalPost": 2,
  "getData": function() {
    return new Data().getDate();
  }
}
```

上面的例子里，这个对象拥有 `posts` 属性。`posts` 的值是一个对象数组，它里面的每一个对象包含键 `title`、`votes` 以及 `comments`。`votes` 属性是一个数值原始值，`comments` 是字符串数组。我们也可以将函数作为值使用，这里 `getData` 的值就是一个函数，这里我们叫它方法。

JSON 与 XML 或者其他数据格式相比，更加灵活、简洁，“JSON: The Fat-Free Alternative to XML”[①]这篇文章对此有详细介绍。MongoDB 使用了一种类似 JSON 的格式叫 BSON[②]，也叫 Binary JSON。第 6 章有更多关于 BSON 的介绍。

3.1.2 AJAX

AJAX 即异步 JavaScript 和 XML，它通过在客户端（浏览器端）使用 XMLHttpRequest 对象与服务器交换数据。虽然 XML 很有名，但目前并不那么常用了，通常使用的是 JSON。这也是现在开发者一般不再说 AJAX 的原因。记住，你也可以同步发 HTTP 请求，但这样做并不好。最常见的同步请求应该是内联 `script` 标签。

---

① http://www.json.org/xml.html

② http://bsonspec.org/

3.1.3 跨域调用

出于安全原因，XMLHTTPRequest 对象最初的实现不允许跨域操作，即：客户端代码和服务器端代码在不同的域中时。我们有很多方法来应对这个问题。

一种常用的方法是使用 JSONP[1]（带填充和前缀的 JSON），本质上它是通过 DOM 操作动态生成的一个 script 标签。script 标签不受同一域限制。JSONP 请求在请求查询字符串中包含回调函数名。比如，`jQuery.ajax()`函数动态生成唯一的函数名，然后附加到请求里（为了方便阅读，下面本来显示为一行的字符串被换行了）：

```
http://graph.facebook.com/search
  ?type=post
  &limit=20
  &q=Gatsby
  &callback=jQuery16207184716751798987_1368412972614&_=1368412984735
```

第二种方法是使用 CORS[2]（Cross-Origin Resource Sharing，跨域资源共享），这是一种更好的解决方案，但是依赖服务器端的控制修改响应头。Chat 示例应用的最终版本将会使用这种方案。

CORS 服务器响应头的示例：

```
Access-Control-Allow-Origin: *
```

更多关于 CORS 的内容：Enable CORS[3]的内容和 HTML5 Rocks Tutorials 网站上的“Using CORS”[4]，可以通过 test-cors.org[5]测试 CORS。

3.2 jQuery

本书中我们都会使用 jQuery（http://jquery.com）进行 DOM 操作、HTTP 请求和 JSONP 调用。jQuery 已成为事实上的标准，因为它的$对象/函数提供了一种非常简便方式，用以通过 ID、类、标签名、属性值、结构或者它们的组合来访问页面中任何 DOM 元素。选择器语法和 CSS 很像，可以使用#作为 ID 选择，. 作为类选择，比如：

```
$("#main").hide();
$("p.large").attr("style","color:red");
$("#main").show().html("<div>new div</div>");
```

下面是一些常用的 jQuery API 函数。

---

① http://en.wikipedia.org/wiki/JSONP

② http://www.w3.org/TR/cors/

③ http://enable-cors.org/resources.html

④ http://www.html5rocks.com/en/tutorials/cors/

⑤ http://client.cors-api.appspot.com/client

- find()[1]：选择由提供的选择器字符串找到的元素。
- hide()[2]：如果一个元素可见，隐藏它。
- show()[3]：如果一个元素隐藏起来了，显示它。
- html()[4]：获取或设置一个元素的 HTML。
- append()[5]：在特定的元素后插入一个元素。
- prepend()[6]：在特定的元素之前插入一个元素。
- on()[7]：添加监听器到元素上。
- off()[8]：删除元素上的监听器。
- css()[9]：获取或设置一个元素的样式属性。
- attr()[10]：获取或设置一个元素的任何属性
- val()[11]：获取或设置一个元素的 value 属性。
- text()[12]：获取一个元素和它包含的元素的文本。
- each()[13]：遍历一组元素。

大多数的 jQuery 函数不是只作用于一个元素上，如果调用它的是一组元素，它会作用于这一组的每一个元素上。这有时候会引发错误，特别是当选择器的范围太广的时候。

jQuery 也拥有大量的插件和库，例如 jQuery UI[14]和 jQuery Mobile[15]，它们提供了大量 UI 和其他功能。

3.3 Twitter Bootstrap

Twitter Bootstrap[16]是一些 CSS、LESS 和 JavaScript 插件的集合，可用于创建用户界面，而不

① http://api.jquery/find
② http://api.jquery/hide
③ http://api.jquery/show
④ http://api.jquery/html
⑤ http://api.jquery/append
⑥ http://api.jquery/prepend
⑦ http://api.jquery/on
⑧ http://api.jquery/off
⑨ http://api.jquery/css
⑩ http://api.jquery/attr
⑪ http://api.jquery/val
⑫ http://api.jquery/text
⑬ http://api.jquery/each
⑭ http://jqueryui.com/
⑮ http://jquerymobile.com/
⑯ http://twitter.github.com/bootstrap/

需要花费时间在诸如圆角按钮、兼容性和响应式等细节上。它可以快速帮你实现想法，构建原型。由于它的可定制性，Twitter Bootstrap 也可非常适合用于正式的项目。源代码是用 LESS[1]写成的，纯 CSS 也可以下载并且正常使用。

下面是一个使用 Twitter Bootstrap 支架的简单例子。项目的文件结构如下所示：

```
/bootstrap
  -index.html
  /css
    -bootstrap.min.css
    ... (other files if needed)
  /img
    glyphicons-halflings.png
    ... (other files if needed)
```

首先我们使用正确的标签创建 index.html：

```
<!DOCTYPE html>
<html lang="en">
  <head>

  </head>
  <body>
  </body>
</html>
```

引入压缩过的 Twitter Bootstrap 的 CSS 文件：

```
<!DOCTYPE html>
<html lang="en">
  <head>
    <link
      type="text/css"
      rel="stylesheet"
      href="css/bootstrap.min.css" />
  </head>
  <body>
  </body>
</html>
```

添加 `container-fluid` 和 `row-fluid` 类：

```
<body >
  <div class="container-fluid">
    <div class="row-fluid">
    </div> <!-- row-fluid -->
  </div> <!-- container-fluid -->
</body>
```

① http://lesscss.org/

Twitter Bootstrap 使用的是 12 列网格。指定某个元素的宽度时可以使用 spanN 类，比如："span1""span2"..."span12"。同样也有用于向右移动的类 offsetN，比如："offset1" "offset2" ... "offset12"。完整的参考信息：http://twitter.github.com/bootstrap/scaffolding.html。

我们为主内容区块使用 span12 和 hero-unit 类：

```
<div class="row-fluid">
 <div class="span12">
    <div id="content">
      <div class="row-fluid">
        <div class="span12">
          <div class="hero-unit">
            <h1>
              Welcome to Super
              Simple Backbone
              Starter Kit
            </h1>
            <p>
              This is your home page.
              To edit it just modify
              <i>index.html</i> file!
            </p>
            <p>
              <a
                class="btn btn-primary btn-large"
                href="http://twitter.github.com/bootstrap"
                target="_blank">
                Learn more
              </a>
            </p>
          </div> <!-- hero-unit -->
        </div> <!-- span12 -->
      </div> <!-- row-fluid -->
    </div> <!-- content -->
  </div> <!-- span12 -->
</div> <!-- row-fluid -->
```

完整的 index.html 源码可以从 rpjs/bootstrap[①]获取：

```
<!DOCTYPE html>
<html lang="en">
<head>
  <link
    type="text/css"
    rel="stylesheet"
    href="css/bootstrap.min.css" />
</head>
  <body>
    <div class="container-fluid">
```

① https://github.com/azat-co/rpjs/tree/master/bootstrap

```
        <div class="row-fluid">
          <div class="span12">
            <div id="content">
              <div class="row-fluid">
                <div class="span12">
                  <div class="hero-unit">
                    <h1>
                      Welcome to Super
                      Simple Backbone
                      Starter Kit
                    </h1>
                    <p>
                      This is your home page.
                      To edit it just modify
                      <i>index.html</i> file!
                    </p>
                    <p>
                      <a
                        class="btn btn-primary btn-large"
                        href="http://twitter.github.com/bootstrap"
                        target="_blank">
                        Learn more
                      </a>
                    </p>
                  </div> <!-- hero-unit -->
                </div> <!-- span12 -->
              </div> <!-- row-fluid -->
            </div> <!-- content -->
          </div> <!-- span12 -->
        </div> <!-- row-fluid -->
      </div> <!-- container-fluid-->
    </body>
    </html>
```

示例的源码可以从 GitHub 公共仓库下载或者拉取，它在 github.com/azat-co/rpjs[①]的 rpjs/bootstrap 目录[②]中。

其他有用的工具——CSS 框架和 CSS 预处理器，如下所示。

- Compass[③]：CSS 框架。
- SASS[④]：类似 LESS，CSS3 扩展。
- Blueprint[⑤]：CSS 框架。

① http://github.com/azat-co/rpjs
② https://github.com/azat-co/rpjs/tree/master/bootstrap
③ http://compass-style.org/
④ http://sass-lang.com/
⑤ http://blueprintcss.org/

- ❑ Foundation[①]：响应式前端框架。
- ❑ Bootswatch[②]：Twitter Bootstrap 自定义主题的集合。
- ❑ WrapBootstrap[③]：Bootstrap 自定义主题的在线商店。

3.4 LESS

LESS 是动态样式语言。有时，特别是在这种情况下，确实是“少（LESS）即是多，多即是少”[④]。

浏览器并不会直接解析 LESS 语法，所以 LESS 源代码必须编译成 CSS 代码，下面是三种可以使用的方式：

(1) 在浏览器里使用 LESS JavaScript library[⑤]；
(2) 在服务器端使用特定的库，即针对 Node.js 可以使用 LESS 模块[⑥]；
(3) 在你自己的电脑上使用诸如 LESS App[⑦]、SimpLESS[⑧]或者类似的应用。

警告

第一种方式在开发的时候是可行的，但是在正式产品中不推荐这么使用。

LESS 支持变量、混入类和操作符，它们可以帮助开发者更快地重复使用 CSS 规则。下面是一个变量的例子。

3.4.1 变量

变量可以减少重复代码，同时因为它们是在一个地方定义的，可以快速改变。大家都知道，在设计阶段经常需要变动值。

LESS 代码：

```
@color: #4D926F;
#header {
  color: @color;
}
```

---

① http://foundation.zurb.com
② http://bootswatch.com
③ https://wrapbootstrap.com/
④ http://en.wikipedia.org/wiki/The_Paradox_of_Choice:_Why_More_Is_Less
⑤ http://lesscss.googlecode.com/files/less-1.3.0.min.js
⑥ https://npmjs.org/package/less
⑦ http://incident57.com/less/
⑧ http://wearekiss.com/simpless

```
h2 {
  color: @color;
}
```

编译后的 CSS 代码：

```
#header {
  color: #4D926F;
}
h2 {
  color: #4D926F;
}
```

3.4.2　混入类（mixin）

下面是混入类的语法，它的行为有点像函数：

```
#header {
  .rounded-corners;
}

#footer {
  .rounded-corners(10px);
}
```

编译后的 CSS 代码：

```
.rounded-corners (@radius: 5px) {
  border-radius: @radius;
  -webkit-border-radius: @radius;
  -moz-border-radius: @radius;
}

#header {
  border-radius: 5px;
  -webkit-border-radius: 5px;
  -moz-border-radius: 5px;
}
#footer {
  border-radius: 10px;
  -webkit-border-radius: 10px;
  -moz-border-radius: 10px;
}
```

混入类可以不使用参数，也可以使用多个参数。

3.4.3　操作符

使用操作符，可以对数字、颜色或者变量进行简单的数学运算。

LESS 中操作符的例子：

```
@the-border: 1px;
@base-color: #111;
@red:        #842210;

#header {
  color: @base-color * 3;
  border-left: @the-border;
  border-right: @the-border * 2;
}
#footer {
  color: @base-color + #003300;
  border-color: desaturate(@red, 10%);
}
```

编译后的 CSS 代码：

```
#header {
  color: #333333;
  border-left: 1px;
  border-right: 2px;
}
#footer {
  color: #114411;
  border-color: #7d2717;
}
```

如你所见，LESS 神奇地增强了原生 CSS 的可重用性。下面是一些可用的在线编译工具。

- LESS2CSS①：很棒的使用 Express.js 构建的 LESS 到 CSS 的转换器。
- lessphp②：在线 demo 编译器。
- Dopefly③：在线 LESS 转换器。

LESS 的其他特性④如下所示：

- 模式匹配
- 嵌套规则
- 函数
- 命名空间
- 作用域
- 注释
- 导入

① http://less2css.org/

② http://leafo.net/lessphp/

③ http://www.dopefly.com/LESS-Converter/less-converter.html

④ http://lesscss.org/#docs

3.5　使用第三方 API（Twitter）和 jQuery 的例子

这个例子完全是为了演示而做。它不是接下来我们要讲解的 Chat 应用的一部分。这个例子的目的是为了演示如何组合 jQuery、JSONP 和 REST API 技术。请阅读代码，但不要尝试运行它，因为现在 Twitter 废弃了 API 1.0 版本。这个应用不会像原来那样运行。如果你执意要自己承担风险运行它，可以按照下面的说明，从 GitHub 下载或者从 PDF 的版本里复制代码。

注意

这个例子是用 Twitter API 1.0 版本构建的，在 Twitter API 1.1 版本下可能无法正常运行，1.1 版本需要用户授权来进行 REST API 调用。可以在 dev.twitter.com 上获取需要的密钥。

在这个例子里，我们使用 jQuery 的`$.ajax()`函数，它的用法如下：

```
var request = $.ajax({
    url: url,
    dataType: "jsonp",
    data: {page:page, ...},
    jsonpCallback: "fetchData"+page,
    type: "GET"
});
```

在上面的代码片段中，我们使用了下面的参数：

- ❑ `url` 是 API 的入口；
- ❑ `dataType` 是我们期望服务器返回的数据类型，比如 `json` 和 `xml`；
- ❑ `data` 是发送到服务器的数据；
- ❑ `jsonpCallback` 是字符串格式的函数名，在请求返回的时候调用；
- ❑ `type` 是 HTTP 请求方法，例如 `GET` 和 `POST`。

更多关于 `ajax()`的参数和例子，请参考 api.jquery.com/jQuery.ajax。

给用户点击事件添加监听器，我们需要使用 jQuery 库里的 `click` 函数。它的用法也非常简单：

```
$("#btn").click(function() {
...
}
```

`$("#btn")`是指向 DOM 中 HTML 元素的一个 jQuery 对象，它的 `id` 是 `btn`。使用了 Twitter Bootstrap 类的按钮的 HTML 代码如下：

```
<input
  type="button"
  class="primary btn"
  id="btn"
  value="Show words in last 1000 tweets"/>
```

为了确保我们使用的 DOM 中的元素都已经在页面中渲染好，需要把对 DOM 修改的代码放在如下的 jQuery 函数中：

```
$(document).ready(function(){
...
}
```

注意

人们通常对动态产生的 HTML 元素有一个误解。它们在创建并注入到 DOM 之前是不存在的。

下面的单页应用可以针对指定的 Twitter 用户所发的推文（200 个单词以内），根据其中单词的使用频率，由高到低依次打印出这些单词。

例如，叫@jack 的用户的推文如下：

```
"hello world"
"hello everyone, and world"
"hi world"
```

返回的结果会是：

```
world
hello
and
hi
everyone.
```

源代码在 rpjs/jquery[1]目录下。它是一个单页应用，只有一个文件 index.html，主要的 JavaScript 算法以下面的方式实现：

```
$(document).ready(function(){
```

我们使用 `document.ready` 推迟执行，直到 DOM 全部加载完成后。

```
$('#btn').click(function() {
```

这给使用了类“btn”的元素添加了点击事件监听器。

```
var username=$('#username').val();
//发请求，执行回调
var url =
 'https://api.twitter.com/1/statuses/user_timeline.json?' +
 'include_entities=true&include_rts=true&screen_name=' +
 username + '&count=1000';
```

实例化变量 `username` 变量，把 `id` 为 `username` 的输入字段里的值赋给它。在接下来的一

---

① https://github.com/azat-co/rpjs/tree/master/jquery

行里，我们把 Twitter REST API 的地址赋值给 url 变量。这个地址返回用户时间轴上的推文。

```
if (username != '') {
  list = []; //存放不重复单词的全局列表
  counter = { };
  var pages = 0;
  getData(url);
}
else {
  alert('Please enter Twitter username')
}
```

为了避免错误的请求，检查 username 变量是否为空。如果用户名正常，使用 getData() 发送请求。这里使用了之后我们将要定义的一个具名函数，这样做是为了防止出现回调嵌套（“臭名昭著”的 pyramid of doom[①]）。

```
  })
});
```

关闭点击回调和 ready 函数代码块。

```
function getData(url) {
  var request = $.ajax({
    url: url,
    dataType: 'jsonp',
    data: {page: 0},
    jsonpCallback: 'fetchData',
    type: 'GET'
  });
}
```

JSONP 获取函数很神奇的（幸好有 jQuery 存在）使用插入 script 标签来进行跨域请求，同时把回调函数名添加到请求查询字符串里。

我们使用 list 数组和 counter 变量来完成这个算法：

```
var list = []; //存放不重复单词的全局列表
var counter = {}; //每个单词重复的次数
```

用来查找和计数的实际函数：

```
function fetchData(m) {
  for (i = 0; i < m.length; i++) {
    var words = m[i].text.split(' ');
    for (j = 0; j < words.length; j++) {
      words[j] = words[j].replace(/\,/g, '');
      //其他代码……
      if (words[j].substring(0, 4) != "http" && words[j] != '') {
        if (list.indexOf(words[j]) < 0) {
```

① http://tritarget.org/blog/2012/11/28/the-pyramid-of-doom-a-JavaScript-style-trap/

```
          list.push(words[j]);
          counter[words[j]] = 1;
        }
        else {
          //给单词计数器加 1
          counter[words[j]]++;
        }
      }
    }
  }
}
```

循环遍历所有的单词，并且使用散列表作为查找表和计数存储。

```
for (i = 0; i < list.length; i++) {
  var max = counter[list[i]];
  var p = 0;
  for (j = i; j < list.length; j++) {
    if (counter[list[i]] < counter[list[j]]) {
      p = list[i];
      list[i] = list[j];
      list[j] = p;
      maxC = i;
    }
  }
}
```

接下来的代码段按重复的次数给单词排序，通过把结果插入到 DOM 中进行展示：

```
  var str = '';
  for (i = 0; i < list.length; i++) {
    str += counter[list[i]] + ': ' + list[i] + '\n';
  }
  $('#log').val(str);
  $('#info').html('Analyzed: ' + list.length +
    'word(s) form ' + m.length +
    'tweet(s).');
}
```

完整的 index.html 源码：

```
$(document).ready(function() {
  $('#btn').click(function() {
    //这里可以替换加载图片
    var username = $('#username').val();
    //发请求，执行回调
    var url =
      'https://api.twitter.com/1/statuses/user_timeline.json?' +
      'include_entities=true&include_rts=true&screen_name=' +
      username + '&count=1000';
    if (username != '') {
      list = []; //存放不重复单词的全局列表
      counter = { };
      var pages = 0;
```

```
      getData(url);
    }
    else {
      alert('Please enter Twitter username')
    }
  })
});
function getData(url) {
  var request = $.ajax({
    url: url,
    dataType: 'jsonp',
    data: {page: 0},
    jsonpCallback: 'fetchData',
    type: 'GET'
  });
}
//ajax 回调函数
var list = []; //存放不重复单词的全局列表
var counter = { };
var pages = 0;

function fetchData(m) {
  for (i = 0; i < m.length; i++) {
    var words = m[i].text.split(' ');
    for (j = 0; j < words.length; j++) {
      words[j] = words[j].replace(/\,/g, '');
   ...
      if (words[j].substring(0, 4) != "http" && words[j] != '') {
        if (list.indexOf(words[j]) < 0) {
          list.push(words[j]);
          counter[words[j]] = 1;
        }
        else {
          //给单词计数器加 1
          counter[words[j]]++;
        }
      }
    }
  }
  //按照重复次数排序
  for (i = 0; i < list.length; i++) {
    var max = counter[list[i]];
    var p = 0;
    for (j = i; j < list.length; j++) {
      if (counter[list[i]] < counter[list[j]]) {
        p = list[i];
        list[i] = list[j];
        list[j] = p;
        maxC = i;
      }
    }
  }
  //打印排序后的结果
  //按照重复的次数很好地打印
```

```
  var str='';
  for (i = 0; i < list.length; i++) {
    str += counter[list[i]] + ': ' + list[i] + '\n';
  }
  $('#log').val(str);
  $('#info').html('Analyzed: ' + list.length +
    'word(s) form ' + m.length +
    'tweet(s).');
}
```

试着在有本地 HTTP 服务器和没有的情况下运行它。**提示**：由于它十分依赖 JSONP，所以在没有 HTTP 服务器的情况下它是不会正常工作的。

注意

这个例子是在 Twitter API 1.0 版本的基础上搭建的，它可能在 Twitter API 1.1 版本下无法正常工作，1.1 版本的 REST API 调用需要用户认证。需要的密钥可以从 dev.twitter.com 获取。如果你觉得必须要有一个可正常运行的版本，请发邮件到 hi@rpjs.co 进行反馈。

3.6 Parse.com

Parse.com 是一个可以用来替换数据库和服务器的服务。它的出现是为了方便移动程序开发。不过，在使用 REST API 和 JavaScript SDK 的情况下，Parse.com 适用于任何 Web 和桌面应用数据存储（及其他功能），因此成为理想的用来快速构建原型的工具。

浏览 Parse.com 并且注册一个免费账户。创建一个应用，复制应用 ID、REST API Key 和 JavaScript Key。我们进入自己在 Parse.com 上的集合的时候需要使用这些密钥。请留心“Data Browser”选项卡，在这里可以看到你的集合和项目。

我们创建一个简单的应用，使用 Parse.com 的 JavaScript SDK 保存数据到集合里。这个应用包含 index.html 和 app.js。下面是项目的结构：

```
/parse
  -index.html
  -app.js
```

这个例子在 GitHub 上的 rpjs/parse[1] 目录，我们鼓励你手动输入代码。从创建 index.html 开始：

```
<html lang="en">
<head>
```

引入谷歌 CDN 上的压缩版 jQuery 库：

① https://github.com/azat-co/rpjs/tree/master/parse

```
<script
  type="text/javascript"
  src=
  "http://ajax.googleapis.com/ajax/libs/jquery/1/jquery.min.js">
</script>
```

从 Parse 的 CDN 上引入 Parse.com 库:

```
<script
  src="http://www.parsecdn.com/js/parse-1.0.14.min.js">
</script>
```

引入 app.js 文件:

```
  <script type="text/javascript" src="app.js"></script>
</head>
<body>
<!-- 我们将在这里做一些事情 -->
</body>
</html>
```

创建 app.js，并且使用`$(document).ready`保证修改的时候 DOM 已经准备好了:

```
$(document).ready(function() {
```

从 Parse.com 控制面板获取并修改 `parseApplicationId` 和 `parseJavaScriptKey` 的值:

```
var parseApplicationId="";
var parseJavaScriptKey="";
```

因为之前已经包含了 Parse JavaScript SDK，现在我们可以使用全局对象 `Parse`。我们使用键初始化一个连接，并且创建一个对 Test 集合的引用:

```
Parse.initialize(parseApplicationId, parseJavaScriptKey);
var Test = Parse.Object.extend("Test");
var test = new Test();
```

下面的示例代码，会保存一个键为 `name` 和 `text` 的对象到 Parse.com 的 `Test` 集合:

```
test.save({
  name: "John",
  text: "hi"}, {
```

为了方便，`save()`方法像 `jQuery.ajax()`函数一样接收回调参数 `success` 和 `error`。为了确认一下，我们只需要看一下浏览器的控制台:

```
    success: function(object) {
      console.log("Parse.com object is saved: " + object);
      //另外也可以使用下面的
      //alert("Parse.com object is saved");
    },
```

知道保存失败的原因也是很重要的：

```
    error: function(object) {
      console.log("Error! Parse.com object is not saved: "+object);
    }
  });
})
```

完整的 index.html 的源码如下：

```
<html lang="en">
<head>
  <script
    type="text/javascript"
    src=
    "http://ajax.googleapis.com/ajax/libs/jquery/1/jquery.min.js">
  </script>
  <script
    src="http://www.parsecdn.com/js/parse-1.0.14.min.js">
  </script>
  <script type="text/javascript" src="app.js"></script>
</head>
<body>
<!-- 我们将在这里做一些事情 -->
</body>
</html>
```

完整的 app.js 的源码如下：

```
$(document).ready(function() {
  var parseApplicationId = "";
  var parseJavaScriptKey = "";
  Parse.initialize(parseApplicationId, parseJavaScriptKey);
  var Test = Parse.Object.extend("Test");
  var test = new Test();
  test.save({
    name: "John",
    text: "hi"}, {
    success: function(object) {
      console.log("Parse.com object is saved: " + object);
      //另外也可以使用下面的
      //alert("Parse.com object is saved");
    },
    error: function(object) {
      console.log("Error! Parse.com object is not saved: " + object);
    }
  });
})
```

警告

这里我们需要使用一个从 Parse.com 控制面板获取的 JavaScript SDK Key。在 jQuery 例子里，我们会使用从相同的网页获取的 REST API Key。如果你从 Parse.com 得到了一个 401 未授权错误，可能是因为你使用了错误的 API 密钥。

一切正常的话，在Parse.com的Data Browser里可以看到`Test`被“John”和“hi”填充。同时，在开发者工具里也可以看到合适的消息。Parse.com自动创建`objectID`和时间戳，这些在我们的Chat应用里会有用。

Parse.com自带一个说明性的Hello World应用，它们在新项目[1]和已有项目[2]的快速开始指南部分。

3.7 使用Parse.com的Chat概述

Chat应用包含一个输入字段、一个消息列表和一个发送按钮。我们需要展示已经存在的消息列表和发送新消息。这里使用Parse.com作为后端，稍后我们会换成结合使用MongoDB的Node.js。

你可以在Parse.com获取免费账户。JavaScript指南可以从https://parse.com/docs/js_guide获取，JavaScript API可以从https://parse.com/docs/js/获取。

注册之后，如果没有应用，到控制面板中创建新应用。复制Application ID和JavaScript SDK key以及REST API Key。稍后我们需要使用它们。这里有一些使用Parse.com的方法。

- REST API：我们将在jQuery的例子里使用这种方法。
- JavaScript SDK：前面的例子里我们使用的是这种方法，在Backbone.js的例子里，我们还会使用这种方法。

REST API是更常见的方式。Parse.com提供了可使用jQuery库中的`$.ajax()`方法请求的入口。这里是一些可用的URL和方法的描述：parse.com/docs/rest。

3.8 使用Parse.com的Chat：REST API和jQuery版本

完整的代码在rpjs/rest[3]目录里，但是我们鼓励你先自己写应用代码。

我们将要使用Parse.com的REST API和jQuery。由于Parse.com支持跨域AJAX请求，所以我们不需要使用JSONP。

注意

当你部署替代Parse.com的后端时，在不同的域名上，要么在前端使用JSONP，要么自定义后端CORS头。本书后面会涉及这个主题。

① https://parse.com/apps/quickstart#js/blank
② https://parse.com/apps/quickstart#js/existing
③ https://github.com/azat-co/rpjs/tree/master/rest

现在的应用结构是这个样子的：

```
index.html
  css/bootstrap.min.css
  css/style.css
  js/app.js
```

让我们来设计 Chat 应用的界面。我们想展示一个以时间为顺序的带有用户名的消息列表。表格可以很好的完成这个工作，而且，当新消息到达的时候动态插入<tr>元素就可以了。

创建一个包含下面内容的 index.html 文件：

- 包含 JS 和 CSS 文件；
- 使用 Twitter Boostrap 进行响应式；
- 一个用来展示消息的表格；
- 用来添加新消息的表单。

我们从 head 标签和依赖开始。我们将包含 CDN 上的 jQuery、本地的 app.js 和本地压缩过的 Twitter Boostrap，以及自定义样式的 style.css：

```
<!DOCTYPE html>
<html lang="en">
  <head>
    <script
      src=
   "https://ajax.googleapis.com/ajax/libs/jquery/1.7.2/jquery.min.js"
      type ="text/javascript"
      language ="javascript" ></script>
    <script src="js/app.js" type="text/javascript"
      language ="javascript" ></script>
    <link href="css/bootstrap.min.css" type="text/css"
      rel="stylesheet" />
    <link href="css/bootstrap-responsive.min.css" type="text/css"
      rel="stylesheet" />
    <link href="css/style.css" type="text/css" rel="stylesheet" />
  </head>
```

body 元素包含 Twitter Boostrap 支架元素，用 container-fluid 和 row-fluid 来定义：

```
<body>
  <div class="container-fluid">
    <div class="row-fluid">
      <h1>Chat with Parse REST API</h1>
```

消息表格是空的，因为我们将会使用 JS 代码动态填充它：

```
  <table class="table table-bordered table-striped">
    <caption>Messages</caption>
    <thead>
      <tr>
```

```
        <th>
          Username
        </th>
        <th>
          Message
        </th>
      </tr>
    </thead>
    <tbody>
      <tr>
        <td colspan="2">No messages</td>
      </tr>
    </tbody>
  </table>
</div>
```

接下来是新消息表单，它的SEND按钮使用了Twitter Boostrap的`btn`和`btn-primary`类：

```
      <div class="row-fluid">
        <form id="new-user">
          <input type="text" name="username"
            placeholder="Username" />
          <input type="text" name="message"
            placeholder="Message" />
          <a id="send" class="btn btn-primary">SEND</a>
        </form>
      </div>
    </div>
  </body>
</html>
```

完整的index.html的源码：

```
<!DOCTYPE html>
<html lang="en">
  <head>
    <script
      src=
   "https://ajax.googleapis.com/ajax/libs/jquery/1.7.2/jquery.min.js"
      type ="text/javascript"
      language ="javascript" ></script>
    <script src="js/app.js" type="text/javascript"
      language ="javascript" ></script>
    <link href="css/bootstrap.min.css" type="text/css"
      rel="stylesheet" />
    <link href="css/bootstrap-responsive.min.css" type="text/css"
      rel="stylesheet" />
    <link href="css/style.css" type="text/css" rel="stylesheet" />
  </head>
  <body>
    <div class="container-fluid">
      <div class="row-fluid">
        <h1>Chat with Parse REST API</h1>
        <table class="table table-bordered table-striped">
```

```
          <caption>Messages</caption>
          <thead>
            <tr>
              <th>
                Username
              </th>
              <th>
                Message
              </th>
            </tr>
          </thead>
          <tbody>
            <tr>
              <td colspan="2">No messages</td>
            </tr>
          </tbody>
        </table>
      </div>
      <div class="row-fluid">
        <form id="new-user">
          <input type="text" name="username"
            placeholder="Username" />
          <input type="text" name="message"
            placeholder="Message" />
          <a id="send" class="btn btn-primary">SEND</a>
        </form>
      </div>
    </div>
  </body>
</html>
```

这个表格将包含我们的消息。表单为新消息提供输入。

现在我们来写三个主要的函数。

(1) `getMessages()`：获取消息的函数。
(2) `updateView()`：渲染消息列表的函数。
(3) `$('#send').click(...)`：触发发送消息的函数。

为了保持简洁，我们把所有的逻辑代码放到 app.js 这个文件里。当然，如果项目变大，需要把代码按功能分成小文件。

使用你自己的值替换下面的值，注意这里使用的是 REST API Key，不是之前例子里使用的 JavaScript SDK Key：

```
var parseID='YOUR_APP_ID';
var parseRestKey='YOUR_REST_API_KEY';
```

把所有的东西打包到 `document.ready` 里，获取消息，定义 SEND 点击事件：

```
$(document).ready(function(){
```

```
  getMessages();
  $("#send").click(function(){
    var username = $('input[name=username]').attr('value');
    var message = $('input[name=message]').attr('value');
    console.log(username)
    console.log('!")
```

提交一个新消息的时候，使用 jQuery.ajax 发送 HTTP 请求：

```
    $.ajax({
      url: 'https://api.parse.com/1/classes/MessageBoard',
      headers: {
        'X-Parse-Application-Id': parseID,
        'X-Parse-REST-API-Key': parseRestKey
      },
      contentType: 'application/json',
      dataType: 'json',
      processData: false,
      data: JSON.stringify({
        'username': username,
        'message': message
      }),
      type: 'POST',
      success: function() {
        console.log('sent');
        getMessages();
      },
      error: function() {
        console.log('error');
      }
    });

  });
})
```

这个方法同样使用 jQuery.ajax 函数获取远程 REST API 服务器的消息列表：

```
function getMessages() {
  $.ajax({
    url: 'https://api.parse.com/1/classes/MessageBoard',
    headers: {
      'X-Parse-Application-Id': parseID,
      'X-Parse-REST-API-Key': parseRestKey
    },
    contentType: 'application/json',
    dataType: 'json',
    type: 'GET',
```

如果请求成功完成（状态是 200 或类似表示），我们调用 updateView 方法：

```
    success: function(data) {
      console.log('get');
      updateView(data);
    },
```

```
      error: function() {
        console.log('error');
      }
    });
  }
```

这个方法把我们从服务器获取的消息渲染：

```
function updateView(messages) {
```

使用 jQuery 选择器.table tbody 选择元素并创建一个引用。接下来，我们清空这个元素的 innerHTML：

```
var table=$('.table tbody');
table.html('');
```

使用 jQuery.each 方法遍历所有消息：

```
$.each(messages.results, function (index, value) {
  var trEl =
```

下面的代码使用编程方式创建 HTML 元素，并且创建这些元素的 jQuery 对象:

```
$('<tr><td>'
  + value.username
  + '</td><td>'
  + value.message +
  '</td></tr>');
```

追加表格 tbody 元素：

```
    table.append(trEl);
  });
  console.log(messages);
}
```

完整的 app.js：

```
var parseID='YOUR_APP_ID';
var parseRestKey='YOUR_REST_API_KEY';

$(document).ready(function(){
  getMessages();
  $("#send").click(function(){
    var username = $('input[name=username]').attr('value');
    var message = $('input[name=message]').attr('value');
    console.log(username)
    console.log('!')
    $.ajax({
      url: 'https://api.parse.com/1/classes/MessageBoard',
      headers: {
        'X-Parse-Application-Id': parseID,
        'X-Parse-REST-API-Key': parseRestKey
      },
      contentType: 'application/json',
```

```
        dataType: 'json',
        processData: false,
        data: JSON.stringify({
          'username': username,
          'message': message
        }),
        type: 'POST',
        success: function() {
          console.log('sent');
          getMessages();
        },
        error: function() {
          console.log('error');
        }
      });

  });
})
function getMessages() {
  $.ajax({
    url: 'https://api.parse.com/1/classes/MessageBoard',
    headers: {
      'X-Parse-Application-Id': parseID,
      'X-Parse-REST-API-Key': parseRestKey
    },
    contentType: 'application/json',
    dataType: 'json',
    type: 'GET',
    success: function(data) {
      console.log('get');
      updateView(data);
    },
    error: function() {
      console.log('error');
    }
  });
}

function updateView(messages) {
  var table=$('.table tbody');
  table.html('');
  $.each(messages.results, function (index, value) {
    var trEl =
      $('<tr><td>'
        + value.username
        + '</td><td>'
        + value.message +
        '</td></tr>');
    table.append(trEl);
  });
  console.log(messages);
}
```

它将会调用 getMessages()方法，然后通过 jQuery 库的$.ajax 发送一个 GET 请求。ajax 方法完整的参数列表可以在 api.jquery.com/jQuery.ajax[1]获取。最重要的参数是 URL、首部和类型参数。

在成功响应时，它会调用 updateView()方法，这个方法清空表格的 tbody 并且使用 jQuery 的$.each 函数（api.jquery.com/jQuery.each[2]）遍历响应结果。点击发送按钮，会发出一个 POST 请求到 Parse.com REST API，接下来，成功响应，再次使用 getMessages 获取所有消息。

下面是使用 jQuery 动态创建 div HTML 元素的一个方法：

```
$("<div>");
```

3.9 推送到 GitHub

为了创建一个 GitHub 仓库，需要先浏览 github.com，登录并且创建新仓库。创建完成后会有一个 SSH 地址，复制它。在终端窗口，切换到你想推送到 GitHub 的项目目录。

(1) 在项目目录的根部初始化一个 Git 库：

```
$ git init
```

(2) 把所有的文件添加到仓库，并开始追踪它们：

```
$ git add .
```

(3) 创建第一条提交：

```
$ git commit -am "inital commit"
```

(4) 添加 GitHub 远程地址：

```
$ git remote add your-github-repo-ssh-url
```

这个地址看起来像这样：

```
$ git remote add origin git@github.com:azat-co/simple-message-board.git
```

(5) 现在可以使用下面的命令把本地 Git 仓库推送到 GitHub 上的远程位置：

```
$ git push origin master
```

(6) 你可以在自己的 github.com 账户和仓库里看到刚才推送的文件。

稍后，当对文件进行修改时，没必要重复上面的所有步骤，只需要执行：

---

① http://api.jquery.com/jQuery.ajax/

② http://api.jquery.com/jQuery.each/

```
$ git add .
$ git commit -am "some message"
$ git push origin master
```

如果没有新的需要跟踪的文件，使用：

```
$ git commit -am "some message"
$ git push origin master
```

提交单独的文件变化，运行：

```
$ git commit filename -m "some message"
$ git push origin master
```

从 Git 仓库里删除文件：

```
$ git rm filename
```

获取更过 Git 命令的信息：

```
$ git --help
```

用 Windows Azure 或者 Heroku 部署应用和推送文件到 GitHub 一样简单。最后的三步（#4-6）将会替换不同的远程地址（URL）和别名。

3.10　部署到 Windows Azure

使用 Git 部署到 Windows Azure。

(1) 浏览 Windows Azure 网站 https://windows.azure.com，使用 Live ID 登录，如果没有站点，创建一个。通过提供用户名和密码（和 Live ID 不同）激活 "Set up Git publishing"。复制你的 URL 到别的地方。

(2) 在你要发布或部署的项目目录的根部初始化一个 Git 库：

```
$ git init
```

(3) 把所有的文件添加到仓库，并开始追踪它们：

```
$ git add .
```

(4) 创建第一条提交：

```
$ git commit -am "inital commit"
```

(5) 添加 Windows Azure 作为 Git 仓库的一个远程地址：

```
$ git remote add azure your-url-for-remote-repository
```

在我这里，这个命令看起来是这样的：

```
$ git remote add

> azure  https://azatazure@azat.scm.azurewebsites.net/azat.git
```

(6) 推送本地 Git 仓库到远程 Windows Azure 仓库，它会部署文件和应用：

```
$ git push azure master
```

和 GitHub 一样，稍后我们更新修改文件的时候，不需要重复前面的几步，因为在项目的根目录已经有了`.git`格式的 Git 仓库描述目录了。

3.11 部署到 Heroku

最大的区别是 Heroku 使用 Cedar Stack，它不支持静态项目，比如：我们之前的测试 Parse.com 应用和使用 Parse.com 版本的 Chat 应用。我们可以使用一个“伪造的”PHP 项目来突破这个限制。在项目目录 index.html 同级创建一个 index.php 文件，我们会把它推荐到 Heroku，它的内容如下：

```
<?php echo file_get_contents('index.html'); ?>
```

为了方便使用，index.php 文件已经包含在 rpjs/rest 里。

这里有一种更简单的使用 Cedar Stack 向 Heroku 上传静态文件的方法，“Static Sites on Heroku Cedar”[1]这篇文章里有详细阐述。为了让 Cedar Static 正常处理你的静态文件，你只需要在项目目录中输入并执行下面的命令：

```
$ touch index.php
$ echo 'php_flag engine off' > .htaccess
```

另外，你也可以使用 Ruby Bamboo 栈。在这种情况下，我们需要有下面的目录结构：

```
-project folder
  -config.ru
  /public
    -index.html
    -/css
    app.js
    ...
```

在 index.html 里引用的 CSS 或者其他静态资源应该使用相对路径，即“css/style.css”。config.ru 文件应该包含下面的代码：

```
use Rack::Static,
  :urls => ["/stylesheets", "/images"],
  :root => "public"

run lambda { |env|
  [
    200,
```

① http://kennethreitz.com/static-sites-on-heroku-cedar.html

```
    {
      'Content-Type'  => 'text/html',
      'Cache-Control' => 'public, max-age=86400'
    },
    File.open('public/index.html', File::RDONLY)
  ]
}
```

更多细节请参考：devcenter.heroku.com/articles/static-sites-on-heroku[①]。

当准备好了所有 Cedar Stack 或者 Bamboo 的支持文件，按照下面的步骤进行。

(1) 如果还没有本地 Git 仓库和.git 目录，创建：

```
$ git init
```

(2) 添加文件：

```
$ git add .
```

(3) 提交文件并更改：

```
$ git commit -m "my first commit"
```

(4) 创建 Heroku Cedar Stack 应用，并且添加远程地址：

```
$ heroku create
```

如果一切正常，它会告诉你远程已经添加，应用已经创建，并且告诉你应用的名字。

(5) 查看远程地址，输入并执行（可选）：

```
$ git remote show
```

(6) 部署代码到 Heroku 使用：

```
$ git push heroku master
```

终端日志会告诉你这次部署的过程。

(7) 在默认的浏览器里打开应用，执行：

```
$ heroku open
```

或者直接浏览你的 app 地址，比如“http://yourappname-NNNN.herokuapp.com”.

(8) 查看 Heroku 上应用的日志：

```
$ heroku logs
```

① https://devcenter.heroku.com/articles/static-sites-on-heroku

使用新的代码更新应用，重复下面的步骤即可：

```
$ git add -A
$ git commit -m "commit for deploy to heroku"
$ git push -f heroku
```

注意

每次使用命令`$ heroku create` 创建一个新的 Heroku 应用的时候，需要为它提供一个新的应用 URL。

3.12 更新和删除消息

按照 REST API 的原则，对一个对象的更新操作最好使用 PUT 方法，删除操作使用 DELETE 方法。这两个都可以通过我们在 GET 和 POST 里使用过的 `jQuery.ajax` 函数简单搞定，只需要把我们希望操作的对象 ID 传入。

第4章

Backbone.js

提要：演示如何从头创建 Backbone.js 应用，在苹果数据库应用示例上使用视图、集合、子视图、模型、事件绑定、AMD 以及 Require.js。

> 代码不是资产，它是债务。写得越多，以后需要维护的也更多。
>
> ——无名氏

4.1 从头开始构建 Backbone.js 应用

我们将使用 Backbone.js 和 MVC 架构创建一个典型的 Hello World 项目。我知道这看上去好像一开始就“杀鸡用牛刀”了，但之后我们会添加更多复杂性，包括模型、子视图和集合。

完整的 Hello World 的源码在 GitHub 上 github.com/azat-co/rpjs/backbone/hello-world[①]中。

依赖

下载下面的库：

- jQuery 1.9 开发源文件[②]；
- Underscore.js 开发源文件[③]；
- Backbone.js 开发源文件[④]。

像这样在 index.html 里引入这些库：

```
<!DOCTYPE>
<html>
<head>
```

① https://github.com/azat-co/rpjs/tree/master/backbone/hello-world

② http://code.jquery.com/jquery-1.9.0.js

③ http://underscorejs.org/underscore.js

④ http://backbonejs.org/backbone.js

```
    <script src="jquery.js"></script>
    <script src="underscore.js"></script>
    <script src="backbone.js"></script>

    <script>
      //TODO 写一些神奇的JS 代码!
    </script>

  </head>
  <body>
  </body>
</html>
```

注意

我们也可以把<script>放到</body>标签之后。这样会改变脚本和余下的HTML加载，在文件较大时可以提升性能。

在<script>标签里定义一个简单的 Backbone.js Router：

```
...
var router = Backbone.Router.extend({
});
...
```

注意

现在，为了尽可能简单，我们把所有的JavaScript代码放到index.html文件里。但在实际的开发中我们并不推荐这样做。稍后我们会重构它。

在 extend 调用里设置指定的 routes 属性：

```
var router = Backbone.Router.extend({
  routes: {
  }
});
```

Backbone.js routes 属性需要下面的格式：'path/:param':'action'，它实现的是当URL是 filename#path/param 时，触发名为 action 的函数（在 Router 对象里定义）。现在我们只添加一个 home 路由：

```
var router = Backbone.Router.extend({
  routes: {
    '': 'home'
  }
});
```

现在我们需要添加一个 home 函数：

```
var router = Backbone.Router.extend({
  routes: {
    '': 'home'
  },
  home:function(){
```

```
    //TODO 渲染 HTML
  }
});
```

一会我们再来完善 home 方法、添加创建和渲染 View 的逻辑。现在我们先定义 homeView：

```
var homeView = Backbone.View.extend({
});
```

看上去很熟悉，是吧？Backbone.js 所有的组件使用的都是相同的语法：extend 函数和传递给它的 JSON 对象参数。

现在很多种方式可以进行，但是最棒的方式还是使用 el 和 template 属性，它们很神奇，即在 Backbone.js 里：

```
var homeView = Backbone.View.extend({
  el:'body',
  template:_.template('Hello World')
});
```

el 是一个保存 jQuery 选择器的字符串，也可以使用'.'作为类和'#'作为 id 名。template 属性被赋值给传入 Hello World 的 Underscore.js 函数 template 运行的结果。

渲染 homeView，我们使用 this.$el，这是一个 jQuery 对象，它指向 el 属性，使用 jQuery.html()函数用 this.template()的结果替换 HTML。下面是我们完整的 Backbone.js View 代码：

```
var homeView = Backbone.View.extend({
  el: 'body',
  template: _.template('Hello World'),
  render: function(){
    this.$el.html(this.template({}));
  }
});
```

现在我们返回到 router，添加两行到 home 函数：

```
var router = Backbone.Router.extend({
  routes: {
    '': 'home'
  },
  initialize: function(){

  },
  home: function(){
    this.homeView = new homeView;
    this.homeView.render();
  }
});
```

第一行创建一个 homeView 对象并且赋值给 router 的 homeView 属性。第二行调用 homeView 对象的 render()方法，触发 Hello World 输出。

最后，启动整体 Backbone 应用，为了保证 DOM 完全加载，用 document-ready 包装器调用 new router：

```
var app;
$(document).ready(function(){
  app = new router;
  Backbone.history.start();
})
```

下面是完整的 index.html 的代码：

```
<!DOCTYPE>
<html>
<head>
  <script src="jquery.js"></script>
  <script src="underscore.js"></script>
  <script src="backbone.js"></script>

  <script>
    var app;
    var router = Backbone.Router.extend({
      routes: {
        '': 'home'
      },
      initialize: function(){
       //一些在对象初始化的时候执行的代码
      },
      home: function(){
        this.homeView = new homeView;
        this.homeView.render();
      }
    });

    var homeView = Backbone.View.extend({
      el: 'body',
      template: _.template('Hello World'),
      render: function(){
        this.$el.html(this.template({}));
      }
    });

    $(document).ready(function(){
      app = new router;
      Backbone.history.start();
    })

  </script>
</head>
<body>
  <div></div>
</body>
</html>
```

在浏览器里打开 index.html 看看它是否正常工作，即“Hello World”消息是否在页面中展现。

4.2 使用集合

这个例子完整的代码在 rpjs/backbone/collections[1]目录下。它在从头开始构建 Backbone.js 应用（4.1 节）的 Hello World 基础上搭建而来（Hello World 可以在 rpjs/backbone/hello-world[2]下载）。

我们需要添加一些模拟用的数据，并且把它们和视图结合。把下面的代码放在 script 标签之后，其他代码之前：

```
var appleData = [
  {
    name: "fuji",
    url: "img/fuji.jpg"
  },
  {
    name: "gala",
    url: "img/gala.jpg"
  }
];
```

这个就是我们的苹果数据库。说得更准确一点儿，这是我们的 REST API 的替代品，它会提供给我们苹果的名字和图像 URL（数据模型）。

注意

在 Backbone.js 里很容易使用你自己的后端 REST API 地址替换模拟数据集，给集合和（或）模型添加一个 `url` 属性，然后调用它们的 `fetch()` 方法。

现在为了使用户体验（UX）更好，我们在 `routes` 对象里添加一个新的路由：

```
...
routes: {
  '': 'home',
  'apples/:appleName': 'loadApple'
},
...
```

这样之后用户在浏览 `index.html#apple/SOMENAME` 的时候可以看到某一个苹果的信息。这个信息由 Backbone `Router` 里定义的 `loadApple` 函数获取和渲染。

```
loadApple: function(appleName){
  this.appleView.render(appleName);
}
```

注意到 `appleName` 变量没？它和我们在 `routes` 里定义的一模一样。这就是在 Backbone.js 里使用查询字符串里参数（比如：`?param=value&q=search`）的方式。

---

① https://github.com/azat-co/rpjs/tree/master/backbone/collections

② https://github.com/azat-co/rpjs/tree/master/backbone/hello-world

现在我们重构一下部分代码，包含创建 Backbone Collection，使它和 appleData 变量绑定，把集合传递给 homeView 及 appleView。简单来说，我们在 Router 的构造函数 initialize 里搞定这些：

```
initialize: function() {
  var apples = new Apples();
  apples.reset(appleData);
  this.homeView = new homeView({collection: apples});
  this.appleView = new appleView({collection: apples});
},
```

这时，我们差不多已经完成了 Router 类，它看起来是这样的：

```
var router = Backbone.Router.extend({
  routes: {
    '': 'home',
    'apples/:appleName': 'loadApple'
  },
  initialize: function() {
    var apples = new Apples();
    apples.reset(appleData);
    this.homeView = new homeView({collection: apples});
    this.appleView = new appleView({collection: apples});
  },
  home: function() {
    this.homeView.render();
  },
  loadApple: function(appleName) {
    this.appleView.render(appleName);
  }
});
```

对 homeView 做一点小改动，让它可以使用整个数据库：

```
var homeView = Backbone.View.extend({
  el: 'body',
  template: _.template('Apple data: <%= data %>'),
  render: function () {
    this.$el.html(this.template({
    data: JSON.stringify(this.collection.models)
  }));
  }
});
```

现在，我们只是把数据以 JSON 字符串的形式展示在浏览器里。这并不是一种用户友好的方式，稍后我们会使用列表和子视图来改进它。

Apple 的 Backbone Collection 非常干净和简单：

```
var Apples = Backbone.Collection.extend({
});
```

注意

Backbone 在集合里使用 fetch()或者 reset()方法时会自动创建一个模型。

Apple 的视图也不复杂，它只包含两个属性 template 和 render。在 template 里我们需要给 figure、img 和 figcaption 标签展示特定的值。Underscore.js 的模板引擎可以很好地处理这个任务：

```
var appleView = Backbone.View.extend({
  template: _.template(
        '<figure>\
           <img src="<%= attributes.url%>"/>\
           <figcaption><%= attributes.name %></figcaption>\
         </figure>'),
...
});
```

为了使一个JavaScript字符串里包含HTML标签同时又保证阅读性，我们可以使用反斜线(\)作为行结束符，也可以使用+来连接字符串，下面是使用后一种方法展示的 appleView：

```
var appleView = Backbone.View.extend({
  template: _.template(
      '<figure>'+
        +'<img src="<%= attributes.url %>"/>'+
        +'<figcaption><%= attributes.name %></figcaption>'+
    +'</figure>'),
...
```

你需要注意<%=和%>符号，它们告诉 Underscore.js 需要展示 attributes 对象的 url、name 属性。

最后，我们给 appleView 添加 render 函数：

```
render: function(appleName){
  var appleModel = this.collection.where({name: appleName})[0];
  var appleHtml = this.template(appleModel);
  $('body').html(appleHtml);
}
```

我们在集合里使用 where()方法和[]选择第一个元素作为模型。现在 render 方法已经可以加载数据和渲染它了。稍后，我们把这两个功能重构分离成不同的方法。

完整的应用在 rpjs/backbone/collections/index.html[①]里，它是这样的：

```
<!DOCTYPE>
<html>
<head>
  <script src="jquery.js"></script>
  <script src="underscore.js"></script>
```

---

① https://github.com/azat-co/rpjs/tree/master/backbone/collections

```
<script src="backbone.js"></script>

<script>
var appleData = [
  {
    name: "fuji",
    url: "img/fuji.jpg"
  },
  {
    name:"gala",
    url: "img/gala.jpg"
  }
 ];
 var app;
 var router = Backbone.Router.extend({
   routes: {
     "": "home",
     "apples/:appleName": "loadApple"
   },
   initialize: function() {
     var apples = new Apples();
     apples.reset(appleData);
     this.homeView = new homeView({ collection: apples });
     this.appleView = new appleView({ collection: apples });
   },
   home: function() {
     this.homeView.render();
   },
   loadApple: function(appleName) {
     this.appleView.render(appleName);
   }
 });
 var homeView = Backbone.View.extend({
   el: 'body',
   template: _.template('Apple data: <%= data %>'),
   render: function() {
     this.$el.html(this.template({
     data: JSON.stringify(this.collection.models)
   }));
   }
   //TODO 子视图
 });

 var Apples = Backbone.Collection.extend({

 });
 var appleView = Backbone.View.extend({
   template: _.template('<figure>\
               <img src="<%= attributes.url %>"/>\
               <figcaption><%= attributes.name %></figcaption>\
             </figure>'),
   //TODO 用加载苹果过程和事件绑定来重写
   render: function(appleName) {
     var appleModel = this.collection.where({
```

4

```
        name: appleName
      })[0];
      var appleHtml = this.template(appleModel);
      $('body').html(appleHtml);
    }
  });
  $(document).ready(function() {
    app = new router;
    Backbone.history.start();
  })

</script>
</head>
<body>
  <div></div>
</body>
</html>
```

在浏览器里打开 collections/index.html 文件。你可以看到从“数据库”里加载的数据：`Appple data:[{"name":"fuji","url":"img/fuji.jpg"},{"name":"gala","url":"img/gala.jpg"}]`。

现在在浏览器里打开 collections/index.html#apples/fuji 或者 collections/index.html#apples/gala。我们期望看到一个带标题的图像。这是一个详细视图，在这个例子里是一个苹果。干得不错！

4.3 事件绑定

实际上，获取数据并不是立即发生的，让我们来重构代码以便模拟它。为了创造更好的用户体验，我们应当展示一个加载图标（Spinner 或 Ajax-loader）告诉用户数据正在加载中。

在 Backbone 里作事件绑定是值得推荐的。如果不这么做，我们必须给请求数据的方法传递一个用于渲染的函数作为回调，以确保渲染方法不会在得到数据之前运行。

因此，当用户浏览详细视图（`apples/:id`），我们只需要调用加载数据的方法。然后，因为设置了正确的监听器，当接收到新数据的时候，视图会自动更新。数据更新的时候，Backbone.js 支持多事件和自定义事件。

改变路由程序里的代码：

```
...
  loadApple: function(appleName){
    this.appleView.loadApple(appleName);
  }
...
```

除了 `appleView` 类，其他的保持不变。我们添加构造函数或者 `initialize` 方法，这是 Backbone.js 框架里一个特殊的字/属性。每个实例初始化的时候会调用这个属性对应的方法，即

var someObj = new SomeObject( )。也可以给 initialize 方法添加额外的参数，就像在视图里做的那样（我们传递了一个以 collection 为键，以 apples Backbone Collection 为值的对象）。更多关于 Backbone.js 的构造函数的信息，请浏览 backbonejs.org/#View-constructor[1]。

```
...
var appleView = Backbone.View.extend({
  initialize: function(){
    //TODO 创建和设置模型，也就是一个苹果
  },
...
```

很好，我们已经有了 initialize 方法。创建相当于单个苹果的模型，并且给它设置正确的监听器。我们将使用两种事件——change 和名为 spinner 的自定义事件。我们将使用 on() 函数，它具备下列属性：on(event, actions, context)。更多相关内容请浏览 backbonejs.org/#Events-on[2]。

```
...
var appleView = Backbone.View.extend({
    this.model = new (Backbone.Model.extend({}));
    this.model.bind('change', this.render, this);
    this.bind('spinner',this.showSpinner, this);
  },
...
```

上面的代码做了两件简单的事情：

(1) 当模型改变时调用 appleView 的 render() 方法；

(2) 当事件 spinner 触发时调用 appleView 的 showSpinner() 方法。

到目前为止，还不错吧？那么 spinner 怎么办呢，用一个简单的 GIF 图标？让我们在 appleView 里创建一个新的属性：

```
...
  templateSpinner: '<img src="img/spinner.gif" width="30"/>',
...
```

还记得路由程序里的 loadApple 调用吗？这是我们在 appleView 里实现函数的方法：

```
...
loadApple:function(appleName){
  this.trigger('spinner');
  //展示加载图
  var view = this;
  //需要在闭包里访问 view
  setTimeout(function(){
  //模拟从远程服务器获取数据消耗的时间
```

① http://backbonejs.org/#View-constructor

② http://backbonejs.org/#Events-on

```
      view.model.set(view.collection.where({
        name:appleName
      })[0].attributes);
    },1000);
  },
  ...
```

第一行触发了 spinner 事件，这个函数是我们必须编写的。

第二行是为了解决作用域问题，我们需要在闭包内使用 appleView。

setTimeout 函数是为了模拟从远程服务器获取响应的时间。在它内部，我们使用 model.set()和 model.attributes（返回模型属性）把选中的模型赋值给当前视图的模型。

现在我们可以移除 render 方法里多余的代码，实现 showSpinner 函数：

```
render: function(appleName){
  var appleHtml = this.template(this.model);
  $('body').html(appleHtml);
},
showSpinner: function(){
  $('body').html(this.templateSpinner);
}
...
```

搞定！用浏览器打开 index.html#apples/gala 或者 index.html#apples/fuji，可以看到加载苹果图像时展示的加载动画。

完整的 index.html 的代码：

```
<!DOCTYPE>
<html>
<head>
  <script src="jquery.js"></script>
  <script src="underscore.js"></script>
  <script src="backbone.js"></script>

  <script>
   var appleData = [
      {
        name: "fuji",
        url: "img/fuji.jpg"
      },
      {
        name: "gala",
        url: "img/gala.jpg"
      }
    ];
    var app;
    var router = Backbone.Router.extend({
      routes: {
        "": 'home',
        "apples/:appleName": 'loadApple'
      },
      initialize: function(){
```

```
    var apples = new Apples();
    apples.reset(appleData);
    this.homeView = new homeView({collection: apples});
    this.appleView = new appleView({collection: apples});
  },
  home: function(){
    this.homeView.render();
  },
  loadApple: function(appleName){
    this.appleView.loadApple(appleName);

  }
});

var homeView = Backbone.View.extend({
  el: 'body',
  template: _.template('Apple data: <%= data %>'),
  render: function(){
    this.$el.html(this.template({
      data: JSON.stringify(this.collection.models)
    }));
  }
  //TODO 子视图
});

var Apples = Backbone.Collection.extend({

});
var appleView = Backbone.View.extend({
  initialize: function(){
    this.model = new (Backbone.Model.extend({}));
    this.model.on('change', this.render, this);
    this.on('spinner',this.showSpinner, this);
  },
  template: _.template('<figure>\
            <img src="<%= attributes.url%>"/>\
            <figcaption><%= attributes.name %></figcaption>\
            </figure>'),
  templateSpinner: '<img src="img/spinner.gif" width="30"/>',

  loadApple:function(appleName){
    this.trigger('spinner');
    var view = this; //需要在闭包里访问 view
    setTimeout(function(){ //模拟从远程服务器获取数据消耗的时间
      view.model.set(view.collection.where({
        name:appleName
      })[0].attributes);
    },1000);

  },

  render: function(appleName){
    var appleHtml = this.template(this.model);
    $('body').html(appleHtml);
```

```
      },
      showSpinner: function(){
        $('body').html(this.templateSpinner);
      }

    });
    $(document).ready(function(){
      app = new router;
      Backbone.history.start();
    })

  </script>
</head>
<body>
  <a href="#apples/fuji">fuji</a>
  <div></div>
</body>
</html>
```

4.4 使用 Underscore.js 视图和子视图

这个例子的代码可以从 rpjs/backbone/subview[①]获取。

子视图也是 Backbone 视图，是在另一个 Backbone 视图里里创建和使用的。子视图对于分离 UI 事件（如点击）和相似模板（比如 apple）是非常棒的一种方式。

使用子视图的情况，可能是表格里的一行，列表里的一项、一个段落、一个新行等。

我们重构主页以展示更好的苹果列表。每一个列表项包含苹果名和带有 onClick 事件的“buy”（购买）链接。我们使用标准 Backbone extend() 函数为单个苹果创建一个子视图：

```
...
var appleItemView = Backbone.View.extend({
  tagName: 'li',
  template: _.template(''
        +'<a href="#apples/<%=name%>" target="_blank">'
      +'<%=name%>'
      +'</a> <a class="add-to-cart" href="#">buy</a>'),
  events: {
    'click .add-to-cart': 'addToCart'
  },
  render: function() {
    this.$el.html(this.template(this.model.attributes));
  },
  addToCart: function(){
    this.model.collection.trigger('addToCart', this.model);
  }
});
...
```

① https://github.com/azat-co/rpjs/tree/master/backbone/subview

现在这个对象拥有 tagName、template、events、render 和 addToCart 等属性和方法。

```
...
tagName : 'li',
...
```

tagName 使 Backbone.js 用指定的标签名创建 HTML 元素，在这个例子是列表里的<li>。这是单个苹果的展示，是列表里的一行。

```
...
template: _.template(''
      +'<a href="#apples/<%=name%>" target="_blank">'
     +'<%=name%>'
     +'</a> <a class="add-to-cart" href="#">buy</a>'),
...
```

模板是一个使用 Underscore.js 函数构建的字符串。它们被<%和%>包裹。<%=简单来说就是指打印值。上面的代码也可以用反斜线的方式表示：

```
...
template: _.template('\
      <a href="#apples/<%=name%>" target="_blank">\
     <%=name%>\
     </a> <a class="add-to-cart" href="#">buy</a>\
     '),
...
```

每一个<li>包含两个链接（<a>），一个指向详细的苹果视图（#apples/:appleName），另一个是 buy 按钮。我们给 buy 按钮添加事件监听器：

```
...
events: {
  'click .add-to-cart': 'addToCart'
},
...
```

它的格式是按下面的规则：

```
事件 + jQuery 选择器: 函数名
```

键和值（冒号右边和左边部分）都是字符串，比如：

```
'click .add-to-cart': 'addToCart'
'click #load-more': 'loadMoreData'
```

这里使用 this.$el 上的 jQuery 方法 html()来展示列表里每一项，this.$el 是根据 tagName 指定的<li>元素：

```
...
render: function() {
  this.$el.html(this.template(this.model.attributes));
},
...
```

addToCart 使用 trigger()函数告诉集合，当前这个特定的模型（苹果）将要被用户购买：

```
...
  addToCart: function(){
    this.model.collection.trigger('addToCart', this.model);
  }
...
```

下面是完整的 appleItemView Backbone View 类代码：

```
...
var appleItemView = Backbone.View.extend({
  tagName: 'li',
  template: _.template(''
        +'<a href="#apples/<%=name%>" target="_blank">'
       +'<%=name%>'
       +'</a> <a class="add-to-cart" href="#">buy</a>'),
  events: {
    'click .add-to-cart': 'addToCart'
  },
  render: function() {
    this.$el.html(this.template(this.model.attributes));
  },
  addToCart: function(){
    this.model.collection.trigger('addToCart', this.model);
  }
});
...
```

简单极了！主视图怎么办呢？它如何渲染每一个项，并且给<li> HTML 元素提供一个包装器<ul>容器呢？我们需要修改增强 homeView。

首先，添加几个简单易懂的 jQuery 选择器字符串作为值的属性给 homeView：

```
...
el: 'body',
listEl: '.apples-list',
cartEl: '.cart-box',
...
```

我们可以在模板里使用上面的属性，或者在 homeView 直接硬编码（稍后会重构）：

```
...
template: _.template('Apple data: \
  <ul class="apples-list">\
  </ul>\
  <div class="cart-box"></div>'),
...
```

创建 homeView 的时候（new homeView）会调用 initialize 方法，在它里面我们渲染模板（使用 html()方法），给集合添加事件监听器。

```
...
  initialize: function() {
    this.$el.html(this.template);
    this.collection.on('addToCart', this.showCart, this);
  },
...
```

在前一节里对于事件绑定我们已经有所讲解。本质上，它会调用 homeView 的 showCart() 方法。在这个方法里，我们给购物车添加 appleName，同时添加一个
元素：

```
...
  showCart: function(appleModel) {
    $(this.cartEl).append(appleModel.attributes.name+'<br/>');
  },
...
```

最后，是我们期待已久的 render()方法，遍历集合里的每一个模型（苹果），为每一个苹果创建 appleItemView，创建<li>，添加到 view.listEl（DOM 里带有 apples-list 类的<ul>元素）：

```
...
render: function(){
  view = this;
  //需要在闭包里访问 view
  this.collection.each(function(apple){
    var appleSubView = new appleItemView({model:apple});
    // 使用苹果模型创建子视图
    appleSubView.render();
    // 使用单个苹果数据渲染模板
    $(view.listEl).append(appleSubView.$el);
    //把渲染好的子视图里的 jQuery 对象添加到苹果列表 DOM 元素里
  });
}
...
```

确保没有在 homeView Backbone View 里遗漏任何事情：

```
...
var homeView = Backbone.View.extend({
  el: 'body',
  listEl: '.apples-list',
  cartEl: '.cart-box',
  template: _.template('Apple data: \
    <ul class="apples-list">\
    </ul>\
    <div class="cart-box"></div>'),
  initialize: function() {
    this.$el.html(this.template);
    this.collection.on('addToCart', this.showCart, this);
  },
  showCart: function(appleModel) {
    $(this.cartEl).append(appleModel.attributes.name+'<br/>');
```

```
  },
  render: function(){
    view = this; //需要在闭包里访问 view
    this.collection.each(function(apple){
      var appleSubView = new appleItemView({model:apple});
      // 使用苹果模型创建子视图
      appleSubView.render();
      // 使用单个苹果数据渲染模板
      $(view.listEl).append(appleSubView.$el);
      //把渲染好的子视图里的 jQuery 对象添加到苹果列表 DOM 元素里
    });
  }
});
...
```

现在可以点击 buy 按钮，购物车里会显示你的选择。查看已经有的苹果，不再需要在浏览器地址栏里输入特定的网址，点击名字，详细视图会在新窗口里展现。

Apple data:

- fuji buy
- gala buy

gala
fuji
fuji
fuji
fuji
fuji
gala
gala
gala
gala
gala

子视图展示的苹果列表

通过子视图，我们重用了模板，并且给每一项（苹果）添加事件。这些事件非常智能的传递模型信息给其他对象：视图及集合。

这个例子里，完整的子视图代码可以从 rpjs/backbone/sub-view/index.html[1]获取。

```
<!DOCTYPE>
<html>
<head>
  <script src="jquery.js"></script>
  <script src="underscore.js"></script>
  <script src="backbone.js"></script>

  <script>
   var appleData = [
      {
        name: "fuji",
```

① https://rpjs/backbone/sub-view/index.html

```
    url: "img/fuji.jpg"
  },
  {
    name: "gala",
    url: "img/gala.jpg"
  }
];
var app;
var router = Backbone.Router.extend({
  routes: {
    "": 'home',
    "apples/:appleName": "loadApple"
  },
  initialize: function(){
    var apples = new Apples();
    apples.reset(appleData);
    this.homeView = new homeView({collection: apples});
    this.appleView = new appleView({collection: apples});
  },
  home: function(){
    this.homeView.render();
  },
  loadApple: function(appleName){
    this.appleView.loadApple(appleName);

  }
});
var appleItemView = Backbone.View.extend({
  tagName: 'li',
  // template: _.template(''
  //     +'<a href="#apples/<%=name%>" target="_blank">'
  //     +'<%=name%>\
  //     +'</a> <a class="add-to-cart" href="#">buy</a>'),
  template: _.template('\
        <a href="#apples/<%=name%>" target="_blank">\
       <%=name%>\
       </a> <a class="add-to-cart" href="#">buy</a>\
       '),

  events: {
    'click .add-to-cart': 'addToCart'
  },
  render: function() {
    this.$el.html(this.template(this.model.attributes));
  },
  addToCart: function(){
    this.model.collection.trigger('addToCart', this.model);
  }
});

var homeView = Backbone.View.extend({
  el: 'body',
  listEl: '.apples-list',
```

4

```
  cartEl: '.cart-box',
  template: _.template('Apple data: \
    <ul class="apples-list">\
    </ul>\
    <div class="cart-box"></div>'),
  initialize: function() {
    this.$el.html(this.template);
    this.collection.on('addToCart', this.showCart, this);
  },
  showCart: function(appleModel) {
    $(this.cartEl).append(appleModel.attributes.name+'<br/>');
  },
  render: function(){
    view = this;
    //需要在闭包里访问view
    this.collection.each(function(apple){
      var appleSubView = new appleItemView({model:apple});
      // 使用苹果模型创建子视图
      appleSubView.render();
      // 使用单个苹果数据渲染模板
      $(view.listEl).append(appleSubView.$el);
      //把渲染好的子视图里的jQuery对象添加到苹果列表DOM元素里
    });
  }
});

var Apples = Backbone.Collection.extend({
});

var appleView = Backbone.View.extend({
  initialize: function(){
    this.model = new (Backbone.Model.extend({}));
    this.model.on('change', this.render, this);
    this.on('spinner',this.showSpinner, this);
  },
  template: _.template('<figure>\
          <img src="<%= attributes.url%>"/>\
          <figcaption><%= attributes.name %></figcaption>\
        </figure>'),
  templateSpinner: '<img src="img/spinner.gif" width="30"/>',
  loadApple:function(appleName){
    this.trigger('spinner');
    var view = this;
    //需要在闭包里访问view
    setTimeout(function(){
    //模拟从远程服务器获取数据消耗的时间
      view.model.set(view.collection.where({
        name:appleName
      })[0].attributes);
    },1000);
  },
  render: function(appleName){
    var appleHtml = this.template(this.model);
    $('body').html(appleHtml);
```

```
    },
    showSpinner: function(){
      $('body').html(this.templateSpinner);
    }
  });

  $(document).ready(function(){
    app = new router;
    Backbone.history.start();
  })

  </script>
</head>
<body>
  <div></div>
</body>
</html>
```

单个项目的链接，比如 collections/index.html#apples/fuji，在浏览器地址栏里输入的时候也是能正常工作的。

4.5　重构

这个时候你可能会好奇使用框架，同时又有大量的类、对象和元素堆积在一个页面上有什么好处。这个是遵循 KISS（Keep it Simple Stupid）原则，尽可能的使事情简单。

应用越大，没有组织的代码带给你的伤痛就越多。我们来把应用拆分成多个文件，每个只包含下面的一种类型：

- 视图
- 模板
- 路由器
- 集合
- 模型

在 index.html 里写入 `script` 标签引入拆分的文件：

```
<script src="apple-item.view.js"></script>
<script src="apple-home.view.js"></script>
<script src="apple.view.js"></script>
<script src="apples.js"></script>
<script src="apple-app.js"></script>
```

文件名不是必须遵循破折号和点分隔，只要可以轻易看出文件的职责就行。

现在我们把对象和类复制到正确的文件里。

index.html 现在看起来非常精简：

```
<!DOCTYPE>
<html>
<head>
  <script src="jquery.js"></script>
  <script src="underscore.js"></script>
  <script src="backbone.js"></script>

  <script src="apple-item.view.js"></script>
  <script src="apple-home.view.js"></script>
  <script src="apple.view.js"></script>
  <script src="apples.js"></script>
  <script src="apple-app.js"></script>

</head>
<body>
  <div></div>
</body>
</html>
```

其他的文件只需要包含与其文件名相符的代码。

apple-item.view.js 文件内容：

```
var appleView = Backbone.View.extend({
  initialize: function() {
    this.model = new(Backbone.Model.extend({}));
    this.model.on('change', this.render, this);
    this.on('spinner', this.showSpinner, this);
  },
  template: _.template('<figure>\
            <img src="<%= attributes.url %>"/>\
            <figcaption><%= attributes.name %></figcaption>\
          </figure>'),
  templateSpinner: '<img src="img/spinner.gif" width="30"/>',

  loadApple: function(appleName) {
    this.trigger('spinner');
    var view = this;
    //需要在闭包里访问view
    setTimeout(function() {
    //模拟从远程服务器获取数据消耗的时间
      view.model.set(view.collection.where({
        name: appleName
      })[0].attributes);
    },1000);

  },

  render: function(appleName) {
    var appleHtml = this.template(this.model);
    $('body').html(appleHtml);
  },
  showSpinner: function() {
    $('body').html(this.templateSpinner);
  }

});
```

apple-home.view.js 拥有 homeView 对象：

```
var homeView = Backbone.View.extend({
  el: 'body',
  listEl: '.apples-list',
  cartEl: '.cart-box',
  template: _.template('Apple data: \
    <ul class="apples-list">\
    </ul>\
    <div class="cart-box"></div>'),
  initialize: function() {
    this.$el.html(this.template);
    this.collection.on('addToCart', this.showCart, this);
  },
  showCart: function(appleModel) {
    $(this.cartEl).append(appleModel.attributes.name+'<br/>');
  },
  render: function(){
    view = this; //需要在闭包里访问 view
    this.collection.each(function(apple){
      var appleSubView = new appleItemView({model:apple});
      // 使用苹果模型创建子视图
      appleSubView.render();
      // 使用单个苹果数据渲染模板
      $(view.listEl).append(appleSubView.$el);
      //把渲染好的子视图里的 jQuery 对象添加到苹果列表 DOM 元素里
    });
  }
});
```

apple.view.js 包含主要的苹果的列表：

```
var appleView = Backbone.View.extend({
  initialize: function(){
    this.model = new (Backbone.Model.extend({}));
    this.model.on('change', this.render, this);
    this.on('spinner',this.showSpinner, this);
  },
  template: _.template('<figure>\
          <img src="<%= attributes.url%>"/>\
          <figcaption><%= attributes.name %></figcaption>\
        </figure>'),
  templateSpinner: '<img src="img/spinner.gif" width="30"/>',
  loadApple:function(appleName){
    this.trigger('spinner');
    var view = this;
    //需要在闭包里访问 view
    setTimeout(function(){
    //模拟从远程服务器获取数据消耗的时间
      view.model.set(view.collection.where({
        name:appleName
      })[0].attributes);
    },1000);
  },
  render: function(appleName){
```

```
    var appleHtml = this.template(this.model);
    $('body').html(appleHtml);
  },
  showSpinner: function(){
    $('body').html(this.templateSpinner);
  }
});
```

apple.js 是一个空集合：

```
var Apples = Backbone.Collection.extend({
});
```

apple-app.js 是应用的主文件，包含数据、路由程序和启动命令：

```
var appleData = [
   {
     name: "fuji",
     url: "img/fuji.jpg"
   },
   {
     name: "gala",
     url: "img/gala.jpg"
   }
];
var app;
var router = Backbone.Router.extend({
  routes: {
    '': 'home',
    'apples/:appleName': 'loadApple'
  },
  initialize: function(){
    var apples = new Apples();
    apples.reset(appleData);
    this.homeView = new homeView({collection: apples});
    this.appleView = new appleView({collection: apples});
  },
  home: function(){
    this.homeView.render();
  },
  loadApple: function(appleName){
    this.appleView.loadApple(appleName);
  }
});
$(document).ready(function(){
  app = new router;
  Backbone.history.start();
})
```

现在打开应用，它看起来应该和之前的子视图示例一模一样。

代码组织已经改善了，但是还没有到完美的地步，因为 HTML 模板还是直接放在了 JavaScript 代码里。这个导致的问题就是设计师和开发者不能在同一个文件上工作，并且所有的改变都需要改变主要的代码。

给 index.html 多添加几个 JS 文件：

```
<script src="apple-item.tpl.js"></script>
<script src="apple-home.tpl.js"></script>
<script src="apple-spinner.tpl.js"></script>
<script src="apple.tpl.js"></script>
```

一般一个 Backbone 视图使用一个模板，在 `appleView` 里，苹果的详细视图里，还有一个加载图（Spinner）和一个加载中的动画（GIF）。

这些文件的内容是一些赋值了字符串的全局变量。稍后在视图里使用 Underscore.js 的 `_.template()`方法时可以使用这些变量。

apple-item.tpl.js 的内容：

```
var appleItemTpl = '\
     <a href="#apples/<%=name%>" target="_blank">\
    <%=name%>\
    </a> <a class="add-to-cart" href="#">buy</a>\
    ';
```

apple-home.tpl.js 的内容：

```
var appleHomeTpl = 'Apple data: \
        <ul class="apples-list">\
        </ul>\
        <div class="cart-box"></div>';
```

apple-spinner.tpl.js：

```
var appleSpinnerTpl = '<img src="img/spinner.gif" width="30"/>';
```

apple.tpl.js 文件：

```
var appleTpl = '<figure>\
                <img src="<%= attributes.url %>"/>\
                <figcaption><%= attributes.name %></figcaption>\
               </figure>';
```

尝试打开这个应用。完全的代码在 rpjs/backbone/refactor[1]目录下。

就像之前的例子一样，我们使用了全局作用域变量（没有使用关键字 `window`）。

警告

当你在全局作用域里使用很多变量时，需要小心。有可能产生冲突和不可预测的问题。例如，如果你写了一个开源库，其他开发人员直接使用方法和属性，而不是接口，那么最后你决定怎么删除和修改那些全局变量？为了防止这样的问题发生，正确的写库和应用方式是使用 JavaScript 闭包[2]。

---

① https://github.com/azat-co/rpjs/tree/master/backbone/refactor

② https://developer.mozilla.org/en-US/docs/JavaScript/Guide/Closures

使用闭包和全局变量模块定义：

```
(function() {
  var apple= function() {
  ...//做一些有意义的事情，比如返回一个苹果对象
  };
  window.Apple = apple;
}())
```

或者当我们需要访问一个 app 对象时，在这个对象上创建依赖：

```
(function() {
  var app = this.app;
  //等价于 window.appliation
  //以防有依赖 (app)
  this.apple = function() {
  ...//返回苹果对象 (类)
  //使用 app 变量
  }
  // 等价于 window.apple = function(){...};
}())
```

如你所见，创建一个匿名函数并且立即执行，把所有的东西包裹在圆括号()里。

4.6　开发时的 AMD 和 Require.js

AMD 允许我们把开发代码组织成模块，管理它们的依赖，异步加载它们。这里有一篇文章解释了为什么 AMD 是个好玩意：WHY AMD?①。

启动你的本地 HTTP 服务器，比如 MAMP②。

index.html 精简成这样：

```
<!DOCTYPE>
<html>
<head>
  <script src="jquery.js"></script>
  <script src="underscore.js"></script>
  <script src="backbone.js"></script>
  <script src="require.js"></script>
  <script src="apple-app.js"></script>
</head>
<body>
  <div></div>
</body>
</html>
```

① http://requirejs.org/docs/whyamd.html

② http://www.mamp.info/en/index.html

只包含库文件和应用的一个文件。这个文件结构如下所示：

```
require([...],function(...){...});
```

说明性更强的展示：

```
require([
  'name-of-the-module',
  ...
  'name-of-the-other-module'
  ],function(referenceToModule, ..., referenceToOtherModule){
  ...// 一些有用的代码
  referenceToModule.someMethod();
});
```

简单来讲，我们通过 require()函数的第一个参数以一个数组形式告诉浏览器我们需要加载的文件列表，把这些文件里的模块传递给匿名回调函数作为参数。在主函数（匿名回调函数）我们可以使用这些模块的引用。因此，我们的 apple-app.js 变形成了：

```
require([
  'apple-item.tpl', // 可使用 shim 插件
  'apple-home.tpl',
  'apple-spinner.tpl',
  'apple.tpl',
  'apple-item.view',
  'apple-home.view',
  'apple.view',
  'apples'
],function(
  appleItemTpl,
  appleHomeTpl,
  appleSpinnerTpl,
  appleTpl,
  appelItemView,
  homeView,
  appleView,
  Apples
  ){
 var appleData = [
    {
      name: "fuji",
      url: "img/fuji.jpg"
    },
    {
      name: "gala",
      url: "img/gala.jpg"
    }
  ];
  var app;
  var router = Backbone.Router.extend({
  //检查是否需要引入
    routes: {
```

```
        '': 'home',
        'apples/:appleName': 'loadApple'
      },
      initialize: function(){
        var apples = new Apples();
        apples.reset(appleData);
        this.homeView = new homeView({collection: apples});
        this.appleView = new appleView({collection: apples});
      },
      home: function(){
        this.homeView.render();
      },
      loadApple: function(appleName){
        this.appleView.loadApple(appleName);

      }
    });

    $(document).ready(function(){
      app = new router;
      Backbone.history.start();
    })
});
```

我们把所有的代码放在 require()的第二个参数的函数里，通过文件名引入它们，通过相应的参数使用依赖关系。现在可以定义模块本身了，我们使用 define()方法：

```
define([...],function(...){...})
```

这和 require()函数相似：依赖以一个字符串数组的形式传入第一个参数。第二个参数是一个接收别的库作为参数的方法，参数的顺序和数组里的模块很重要：

```
define(['name-of-the-module'],function(nameOfModule){
  var b = nameOfModule.render();
  return b;
})
```

注意

不必给文件名添加.js，Require.js 会自动添加。shim 插件可以用来导入文本文件作为模板。

现在开始处理模板，把它们转换成 Require.js 模块。

新的 apple-item.tpl.js 文件：

```
define(function() {
  return '\
           <a href="#apples/<%=name%>" target="_blank">\
          <%=name%>\
          </a> <a class="add-to-cart" href="#">buy</a>\
          '
});
```

apple-home.tpl.js 文件：

```
define(function(){
  return 'Apple data: \
        <ul class="apples-list">\
        </ul>\
        <div class="cart-box"></div>';
});
```

apple-spinner.tpl.js 文件：

```
define(function(){
  return '<img src="img/spinner.gif" width="30"/>';
});
```

apple.tpl.js 文件：

```
define(function(){
  return '<figure>\
          <img src="<%= attributes.url %>"/>\
          <figcaption><%= attributes.name %></figcaption>\
         </figure>';
});
```

apple-item.view.js 文件：

```
define(function() {
  return '\
             <a href="#apples/<%=name%>" target="_blank">\
            <%=name%>\
            </a> <a class="add-to-cart" href="#">buy</a>\
            '
});
```

在 apple-home.view.js 文件里，需要声明它依赖 apple-home.tpl 和 apple-item.view.js 文件：

```
define(['apple-home.tpl', 'apple-item.view'], function(
  appleHomeTpl,
  appleItemView) {
return  Backbone.View.extend({
      el: 'body',
      listEl: '.apples-list',
      cartEl: '.cart-box',
      template: _.template(appleHomeTpl),
      initialize: function() {
        this.$el.html(this.template);
        this.collection.on('addToCart', this.showCart, this);
      },
      showCart: function(appleModel) {
        $(this.cartEl).append(appleModel.attributes.name + '<br/>');
      },
      render: function() {
        view = this; //需要在闭包里访问view
```

4

```
          this.collection.each(function(apple) {
            var appleSubView = new appleItemView({ model: apple });
            // 使用苹果模型创建子视图
            appleSubView.render();
            // 使用单个苹果数据渲染模板
            $(view.listEl).append(appleSubView.$el);
            //把渲染好的子视图里的 jQuery 对象添加到苹果列表 DOM 元素里
          });
        }
      });
})
```

apple.view.js 依赖两个模板：

```
define([
  'apple.tpl',
  'apple-spinner.tpl'
], function(appleTpl, appleSpinnerTpl) {
    return Backbone.View.extend({
      initialize: function() {
        this.model = new(Backbone.Model.extend({}));
        this.model.on('change', this.render, this);
        this.on('spinner', this.showSpinner, this);
      },
      template: _.template(appleTpl),
      templateSpinner: appleSpinnerTpl,
      loadApple: function(appleName) {
        this.trigger('spinner');
        var view = this;
        //需要在闭包里访问 view
        setTimeout(function() {
        //模拟从远程服务器获取数据消耗的时间
          view.model.set(view.collection.where({
            name: appleName
          })[0].attributes);
        }, 1000);
      },
      render: function(appleName) {
        var appleHtml = this.template(this.model);
        $('body').html(appleHtml);
      },
      showSpinner: function() {
        $('body').html(this.templateSpinner);
      }
  });
});
```

apple.js 文件：

```
define(function(){
    return Backbone.Collection.extend({})
});
```

我希望你已经看到这个模式了。所有的代码基于逻辑分离到不同的文件里（比如视图类、集合类、模板）。主文件使用 `require()` 方法加载所有的依赖。如果在非主文件里需要加载一些文件，在 `define()` 里请求它们。一般在模块里我们返回一个对象，例如在模板里我们返回一个字符串，在视图模块我们返回 Backbone View 类。

尝试打开 rpjs/backbone/amd[①]里例子。展现上不会有任何变化。如果打开开发者工具的 Network（网络）面板，可以看到文件加载的不同。旧的 rpjs/backbone/refactor/index.html[②]串行加载 JS 脚本，新的 rpjs/backbone/amd/index.html[③]并行加载 JS 脚本。

老的 rpjs/backbone/refactor/index.html 文件

① https://github.com/azat-co/rpjs/tree/master/backbone/amd

② https://github.com/azat-co/rpjs/blob/master/backbone/refactor/index.html

③ https://github.com/azat-co/rpjs/blob/master/backbone/amd/index.html

Apple data:

- fuji buy
- gala buy

Elements Resources Network Sources Timeline Profiles Audits Console

Name / Path	Method	Status / Text	Type	Initiator	Size / Content	Time / Latency	Timeline
index.html /General%20Assembly/ga-backbone/.	GET	304 Not Modified	text/html	Other	172 B 266 B	4 ms 3 ms	
jquery.js /General%20Assembly/ga-backbone/.	GET	304 Not Modified	applicatio...	index.html:4 Parser	173 B 261 KB	10 ms 7 ms	
underscore.js /General%20Assembly/ga-backbone/.	GET	304 Not Modified	applicatio...	index.html:4 Parser	172 B 40.4 KB	36 ms 10 ms	
backbone.js /General%20Assembly/ga-backbone/.	GET	304 Not Modified	applicatio...	index.html:4 Parser	172 B 54.5 KB	56 ms 13 ms	
require.js /General%20Assembly/ga-backbone/.	GET	304 Not Modified	applicatio...	index.html:4 Parser	174 B 79.2 KB	60 ms 12 ms	
apple-app.js /General%20Assembly/ga-backbone/.	GET	304 Not Modified	applicatio...	index.html:4 Parser	172 B 1.2 KB	67 ms 12 ms	
nudge-icon-arrow-up.png pioclpoplcdbaefihamjohnefbikjilc/ima	GET	200 OK	image/png	Preview.js:41 Script	(from cac...	118 ms 118 ms	
nudge-icon-arrow-down.png pioclpoplcdbaefihamjohnefbikjilc/ima	GET	200 OK	image/png	Preview.js:41 Script	(from cac...	118 ms 118 ms	
nudge-icon-arrow-lr.png pioclpoplcdbaefihamjohnefbikjilc/ima	GET	200 OK	image/png	Preview.js:41 Script	(from cac...	118 ms 118 ms	
nudge-icon-return.png pioclpoplcdbaefihamjohnefbikjilc/ima	GET	200 OK	image/png	Preview.js:41 Script	(from cac...	118 ms 118 ms	
apple-spinner.tpl.js /General%20Assembly/ga-backbone/.	GET	304 Not Modified	applicatio...	require.js:1854 Script	170 B 74 B	16 ms 12 ms	
apple-item.tpl.js /General%20Assembly/ga-backbone/.	GET	304 Not Modified	applicatio...	require.js:1854 Script	170 B 200 B	14 ms 7 ms	
apple-item.view.js /General%20Assembly/ga-backbone/.	GET	304 Not Modified	applicatio...	require.js:1854 Script	172 B 616 B	17 ms 14 ms	
apple-home.view.js /General%20Assembly/ga-backbone/.	GET	304 Not Modified	applicatio...	require.js:1854 Script	171 B 951 B	16 ms 12 ms	
apple.tpl.js	GET	304	applicatio	require.js:1854	171 B	14 ms	

Timeline: 159 ms 238 ms 318 ms 397 ms 477 ms

Documents Stylesheets Images Scripts XHR Fonts WebSockets Other

新的 rpjs/backbone/amd/index.html 文件

Require.js 可以通过在 `requirejs.config()` HTML 页面顶部调用定义配置。更多信息可以查阅：requirejs.org/docs/api.html#config[①]。

我们来给这个例子添加时间戳，时间戳会添加到每个 URL 之后防止浏览器缓存。这种方式在开发阶段非常棒，但是在线上产品中是一个非常糟糕的主意（线上产品就尽可能使用浏览器缓存）。

在 apple-app.js 文件开始的地方添加：

```
requirejs.config({
  urlArgs: "bust=" + (new Date()).getTime()
});
require([
...
```

① http://requirejs.org/docs/api.html#config

Apple data:

- fuji buy
- gala buy

Elements Resources Network Sources Timeline Profiles Audits Console

Name / Path	Method	Status / Text	Type	Initiator	Size / Content	Time / Latency
nudge-icon-arrow-up.png pioclpoplcdbaefihamjohnefbikjilc/ima	GET	200 OK	image/png	Preview.js:41 Script	(from cac...	127 ms 127 ms
nudge-icon-arrow-down.png pioclpoplcdbaefihamjohnefbikjilc/ima	GET	200 OK	image/png	Preview.js:41 Script	(from cac...	127 ms 127 ms
nudge-icon-arrow-lr.png pioclpoplcdbaefihamjohnefbikjilc/ima	GET	200 OK	image/png	Preview.js:41 Script	(from cac...	127 ms 127 ms
nudge-icon-return.png pioclpoplcdbaefihamjohnefbikjilc/ima	GET	200 OK	image/png	Preview.js:41 Script	(from cac...	127 ms 127 ms
apple-item.tpl.js?bust=136358857: /General%20Assembly/ga-backbone/.	GET	200 OK	applicatio...	require.js:1854 Script	(from cac...	23 ms 23 ms
apple-home.tpl.js?bust=136358857 /General%20Assembly/ga-backbone/.	GET	200 OK	applicatio...	require.js:1854 Script	422 B 135 B	19 ms 9 ms
apple-spinner.tpl.js?bust=1363588 /General%20Assembly/ga-backbone/.	GET	200 OK	applicatio...	require.js:1854 Script	360 B 74 B	20 ms 11 ms
apple.tpl.js?bust=1363588572503 /General%20Assembly/ga-backbone/.	GET	200 OK	applicatio...	require.js:1854 Script	542 B 254 B	27 ms 14 ms
apple-home.view.js?bust=1363588: /General%20Assembly/ga-backbone/.	GET	200 OK	applicatio...	require.js:1854 Script	1.2 KB 951 B	30 ms 27 ms
apple-item.view.js?bust=13635885 /General%20Assembly/ga-backbone/.	GET	200 OK	applicatio...	require.js:1854 Script	904 B 616 B	27 ms 14 ms
apple.view.js?bust=136358857250: /General%20Assembly/ga-backbone/.	GET	200 OK	applicatio...	require.js:1854 Script	1.2 KB 952 B	30 ms 27 ms
apples.js?bust=1363588572503 /General%20Assembly/ga-backbone/.	GET	200 OK	applicatio...	require.js:1854 Script	355 B 69 B	26 ms 22 ms
data:image/png;base...	GET	Success	image/png	10101 gc-mini-butt Script	0 B 487 B	0 ms 0
storify-common.css oonhlodhpiagekajhhfimfgnagjnnop/c	GET	200 OK	text/css	jquery.js:3 Script	(from cac...	88 ms 88 ms

20 requests | 5.8 KB transferred | 727 ms (onload: 743 ms, DOMContentLoaded: 451 ms)

All | Documents Stylesheets Images Scripts XHR Fonts WebSockets Other

添加了时间戳的网络面板

可以看到所有的请求现在状态码都是 200 而不是 304（没有修改）。

4.7 生产环境里的 Require.js

使用 NPM（Node Package Manager）安装 requirejs 库，在你的项目目录里，或者在终端里运行这个命令：

```
$ npm install requirejs
```

或者添加-g 作为全局安装：

```
$ npm install -g requirejs
```

创建 app.build.js 文件：

```
({
    appDir: "./js",
    baseUrl: "./",
    dir: "build",
    modules: [
        {
            name: "apple-app"
        }
    ]
})
```

把这个脚本文件移到 `js` 目录，对应的是 `appDir` 属性。生成的文件会放在 build 目录，对应的是 `dir` 参数。更多关于生成文件，查看这个使用了很多功能带有注释的文件：https://github.com/jrburke/r.js/blob/master/build/example.build.js。

现在所有的准备就绪，我们可以生成一个大 JS 文件，它包含了所有的依赖和模块：

```
$ r.js -o app.build.js
```

或者：

```
$ node_modules/requirejs/bin/r.js -o app.build.js
```

这会显示出 r.js 处理的文件列表。

```
Uglifying file: /Users/azat/Documents/Development/General Assembly/ga-backbone/r/build/apple-app.js
Uglifying file: /Users/azat/Documents/Development/General Assembly/ga-backbone/r/build/apple-home.tpl.js
Uglifying file: /Users/azat/Documents/Development/General Assembly/ga-backbone/r/build/apple-home.view.js
Uglifying file: /Users/azat/Documents/Development/General Assembly/ga-backbone/r/build/apple-item.tpl.js
Uglifying file: /Users/azat/Documents/Development/General Assembly/ga-backbone/r/build/apple-item.view.js
Uglifying file: /Users/azat/Documents/Development/General Assembly/ga-backbone/r/build/apple-spinner.tpl.js
Uglifying file: /Users/azat/Documents/Development/General Assembly/ga-backbone/r/build/apple.tpl.js
Uglifying file: /Users/azat/Documents/Development/General Assembly/ga-backbone/r/build/apple.view.js
Uglifying file: /Users/azat/Documents/Development/General Assembly/ga-backbone/r/build/apples.js
Uglifying file: /Users/azat/Documents/Development/General Assembly/ga-backbone/r/build/backbone.js
Uglifying file: /Users/azat/Documents/Development/General Assembly/ga-backbone/r/build/jquery.js
toTransport skipping /Users/azat/Documents/Development/General Assembly/ga-backbone/r/build/node_modules/.bin/r.js: Error: Line 1: Unexpe
cted token ILLEGAL
Error: Cannot parse file: /Users/azat/Documents/Development/General Assembly/ga-backbone/r/build/node_modules/.bin/r.js for comments. Ski
pping it. Error is:
Error: Line 1: Unexpected token ILLEGAL
toTransport skipping /Users/azat/Documents/Development/General Assembly/ga-backbone/r/build/node_modules/requirejs/bin/r.js: Error: Line
1: Unexpected token ILLEGAL
Error: Cannot parse file: /Users/azat/Documents/Development/General Assembly/ga-backbone/r/build/node_modules/requirejs/bin/r.js for comm
ents. Skipping it. Error is:
Error: Line 1: Unexpected token ILLEGAL
Uglifying file: /Users/azat/Documents/Development/General Assembly/ga-backbone/r/build/node_modules/requirejs/require.js
Uglifying file: /Users/azat/Documents/Development/General Assembly/ga-backbone/r/build/require.js
Uglifying file: /Users/azat/Documents/Development/General Assembly/ga-backbone/r/build/underscore.js

apple-app.js
----------------
apple-item.tpl.js
apple-home.tpl.js
apple-spinner.tpl.js
apple.tpl.js
apple-item.view.js
apple-home.view.js
apple.view.js
apples.js
apple-app.js

r git:(master) ✗ $ node_modules/requirejs/bin/r.js -o app.build.js
```

r.js 处理的文件列表

从 build 目录打开 index.html，查看 Network 面板，你可以看到只有一个请求文件。

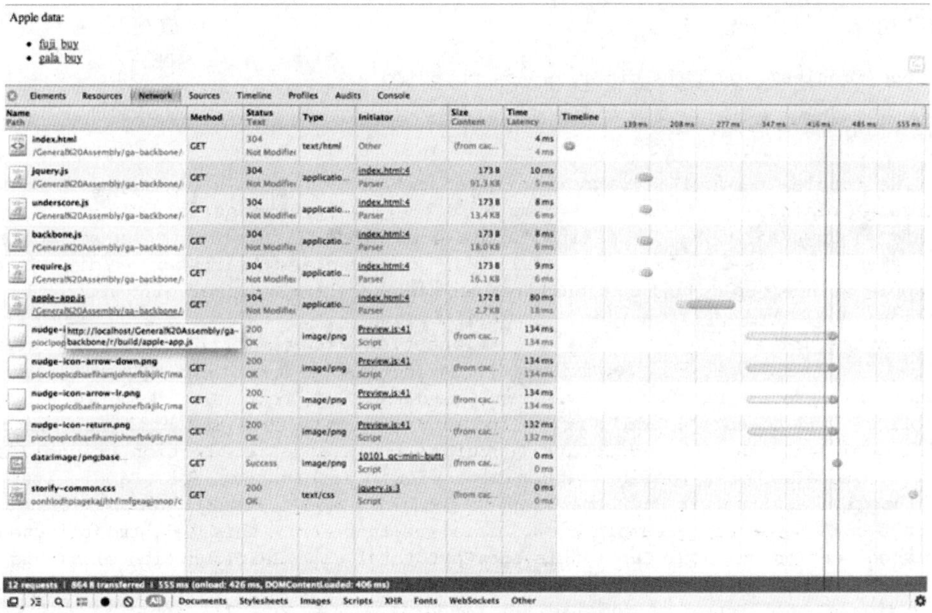

通过只加载一个文件增强性能

更多信息可以查阅官方的 r.js 文档：requirejs.org/docs/optimization.html①。

示例代码在 rpjs/backbone/r②和 rpjs/backbone/r/build③目录里。

为了减小文件大小，我们压缩文件，使用 Uglify2④模块完成这个任务。使用 NPM 来安装它：

```
$ npm install uglify-js
```

更新 app.build.js 文件，添加一个 `optimize:"uglify2"`属性：

```
({
    appDir: "./js",
    baseUrl: "./",
    dir: "build",
    optimize: "uglify2",
    modules: [
        {
            name: "apple-app"
        }
    ]
})
```

① http://requirejs.org/docs/optimization.html

② https://github.com/azat-co/rpjs/tree/master/backbone/r

③ https://github.com/azat-co/rpjs/tree/master/backbone/r/build

④ https://github.com/mishoo/UglifyJS2

运行 r.js:

```
$ node_modules/requirejs/bin/r.js -o app.build.js
```

现在得到的将会是:

```
define("apple-item.tpl",[],function(){return'           <a href="#apples/<%=name%>" target="_blank">            <%=name%>            </a> <a class="add-to-cart" href="#">buy</a>        '}),define("apple-home.tpl",[],function(){return'Apple data: <ul class="apples-list">         </ul>        <div class="cart-box"></div>'}),define("apple-spinner.tpl",[],function(){return'<img src="img/spinner.gif" width="30"/>'}),define("apple.tpl",[],function(){return'<figure>                   <img src="<%= attributes.url%>"/>                   <figcaption><%= attributes.name %></figcaption>                   </figure>'}),define("apple-item.view", ["apple-item.tpl"],function(e){return Backbone.View.extend({tagName:"li", template:_.template(e),events:{"click  .add-to-cart":"addToCart"},render:function(){this.$el.html(this.template(this.model.attributes))},addToCart:function(){this.model.collection.trigger("addToCart",this.model)}})}),define("apple-home.view",["apple-home.tpl","apple-item.view"], function(e,t){return Backbone.View.extend({el: "body",listEl:".apples-list",cartEl:".cart-box", template:_.template(e),initialize:function(){this.$el.html(this.template),this.collection.on("addToCart",this.showCart,this)},showCart:function(e){$(this.cartEl).append(e.attributes.name+"<br/>")},render:function(){view=this,this.collection.each(function(e){var i=new t({model:e});i.render(),$(view.listEl).append(i.$el)})}})}),define ("apple.view",["apple.tpl","apple-spinner.tpl"],function(e,t){return Backbone.View.extend({initialize:function(){this.model=new(Backbone.Model.extend({})),this.model.on("change",this.render,this),this.on("spinner",this.showSpinner,this)},template:_.template(e),templateSpinner:t,loadApple:function(e){this.trigger("spinner");var t=this;setTimeout(function(){t.model.set(t.collection.where({name:e})[0].attributes)},1e3)},render:function(){var e=this.template(this.model);$("body").html(e)},showSpinner:function(){$("body").html(this.templateSpinner)}})}),define("apples",[],function(){return Backbone.Collection.extend({})}),requirejs.config({urlArgs:"bust="+(new Date).getTime()}),require(["apple-item.tpl","apple-home.tpl","apple-spinner.tpl","apple.tpl","apple-item.view","apple-home.view","apple.view","apples"],function(e,t,i,n,a,l,p,o){var  r,s=[{name:"fuji",url:"img/fuji.jpg"},{name:"gala",url:"img/gala.jpg"}],c=Backbone.Router.extend({routes:{"":"home","apples/:appleName":"loadApple"},initialize:function(){var  e=new  o;e.reset(s),this.homeView=new  l({collection:e}),this.appleView=new  p({collection:e})},home:function(){this.homeView.render()},loadApple:function(e){this.appleView.loadApple(e)}});$(document).ready(function(){r=new c,Backbone.history.start()})}),define("apple-app",function(){});
```

注意

为了展示 Uglify2 是怎么工作的，这个文件没有格式化。在没有续行符的情况下，这段代码是一行。同时需要注意，变量和对象名也压缩了。

4.8　简单好用的 Backbone 脚手架工具

快速开始 Backbone.js 开发，可以考虑使用 Super Simple Backbone Starter Kit[1]或者类似的项

① https://github.com/azat-co/super-simple-backbone-starter-kit

目：

- Backbone Boilerplate[①]；
- Sample App with Backbone.js and Twitter Bootstrap[②]；
- 更多有关 Backbone.js 的教程可以浏览 github.com/documentcloud/backbone/wiki/Tutorials%2C-blog-posts-and-example-sites[③]获取。

4

① http://backboneboilerplate.com/

② http://coenraets.org/blog/2012/02/sample-app-with-backbone-js-and-twitter-bootstrap/

③ https://github.com/documentcloud/backbone/wiki/Tutorials%2C-blog-posts-and-example-sites

第 5 章

Backbone.js 和 Parse.com

提要：使用 Parse.com 和它的 JavaScript SDK 在修改后的 Chat 应用上试验 Backbone.js；建议添加的功能列表。

> "Java 和 JavaScript 的关系，就好比'雷锋'和'雷峰塔'的关系。"
>
> ——克里斯·海尔曼[①]

如果你已经写过一些复杂的客户端程序，会发现维护 JavaScript 回调函数与 UI 事件混杂在一起的代码很有挑战性。Backbone.js 提供了一个轻量且强大的方式，用于把逻辑代码组织成 MVC 结构。它同样有一些特别棒的功能，如 URL 路由选择、REST API 支持、事件监听器与触发器。更多关于从头开始构建 Backbone.js 应用的信息和详细例子，请参考第 4 章。

可以从 backbonejs.org[②]下载 Backbone.js 库。然后，像其他 JavaScript 一样在 HTML 页面的顶端或者正文中引入它，你就可以使用 Backbone 类。例如，创建路由程序：

```
var ApplicationRouter = Backbone.Router.extend({
  routes: {
    "": "home",
    "signup": "signup",
    "*actions": "home"
  },
  initialize: function () {
    this.headerView = new HeaderView();
    this.headerView.render();
    this.footerView = new FooterView();
    this.footerView.render();
  },
  home: function () {
    this.homeView = new HomeView();
    this.homeView.render();
  },
  signup: function () {
    ...
  }
});
```

① http://christianheilmann.com/

② http://backbonejs.org

视图、模型和集合以同样的方式创建：

```
HeaderView = Backbone.View.extend({
  el: "#header",
  template: '<div>...</div>',
  events: {
    "click #save": "saveMessage"
  },
  initialize: function () {
    this.collection = new Collection();
    this.collection.bind("update", this.render, this);
    },
  saveMessage: function () {
    ...
  },
  render: function () {
    $(this.el).html(_.template(this.template));
  }
});

Model = Backbone.Model.extend({
  url: "/api/item"
  ...
});

Collection = Backbone.Collection.extend({
  ...
});
```

更多关于 Backbone.js 的细节，请参考第 4 章。

5.1 使用 Parse.com 的 Chat：JavaScript SDK 和 Backbone.js 版本

显而易见，如果继续添加越来越多的像“DELETE”“UPDATE”这样的按钮，我们的系统里异步回调会变得越来越复杂。同时，我们必须知道什么时候更新视图，即消息列表，这取决于数据是否改变。使用 Backbone.js 的 MVC 框架更易管理与维护复杂应用。

如果你感觉前一个例子比较轻松，让我们在此基础之上，使用 Backbone.js 框架继续搭建它。我们将一步一步的使用 Backbone.js 和 Parse.com 的 JavaScript SDK 创建 Chat 应用。如果你感觉已经很熟悉了，可以下载 Backbone.js 的脚手架工具：github.com/azat-co/super-simple-backbone-starter-kit[①]。集成 Backbone.js 后，可以直接把用户操作的异步更新集合绑定起来。

这个应用在 rpjs/sdk[②]可以获取，但我们还是建议你从头开始写代码，示例代码仅仅作为参考。

下面是使用了 Parse.com 的 Chat 的目录结构：

---

① https://github.com/azat-co/super-simple-backbone-starter-kit

② https://github.com/azat-co/rpjs/tree/master/sdk

```
/sdk
  -index.html
  -home.html
  -footer.html
  -header.html
  -app.js
  /css
    -bootstrap-responsive.css
    -bootstrap-responsive.min.css
    -bootstrap.css
    -bootstrap.min.css
  /img
    -glyphicons-halflings-white.png
    -glyphicons-halflings.png
  /js
    -bootstrap.js
    -bootstrap.min.js
  /libs
    -require.min.js
    -text.js
```

创建一个文件夹，在这个文件佚里创建 index.html，输入下面的内容：

```
<!DOCTYPE html>
<html lang="en">
  <head>
  ...
  </head>
  <body>
  ...
  </body>
</html>
```

下载必须的库或者使用 Google API 上的链接。现在在 head 元素里引入 JavaScript 库代码和 Twitter Bootstrap 样式表，同时写上一些重要但是不是必需的 meta 元素。

```
<head>
  <meta charset="utf-8"/>
  <title></title>
  <meta name="description" content=""/>
  <meta name="author" content=""/>
```

为响应式做好准备：

```
<meta name="viewport"
  content="width=device-width, initial-scale=1.0" />
```

从 Google API 引入 jQuery：

```
  <script type="text/javascript"
src="http://ajax.googleapis.com/ajax/libs/jquery/1.7.2/jquery.min.js">
  </script>
```

引入 Twitter Bootstrap 插件：

```
<script type="text/javascript" src="js/bootstrap.min.js"></script>
```

从 Parse.com CDN 引入 Parse JavaScript SDK：

```
<script type="text/javascript"
  src="http://www.parsecdn.com/js/parse-1.0.5.min.js">
</script>
```

引入 Twitter Bootstrap 的 CSS 样式文件：

```
<link type="text/css"
  rel="stylesheet"
  href="css/bootstrap.min.css" />
<link type="text/css"
  rel="stylesheet"
  href="css/bootstrap-responsive.min.css" />
```

引入我们的 JS 应用：

```
<script type="text/javascript" src="app.js"></script>
</head>
```

给 body 元素添上 Twitter Bootstrap 的支架内容（详情请参考第 1 章）：

```
<body>
<div class="container-fluid">
  <div class="row-fluid">
    <div class="span12">
      <div id="header">
      </div>
    </div>
  </div>
  <div class="row-fluid">
    <div class="span12">
      <div id="content">
      </div>
    </div>
  </div>
  <div class="row-fluid">
    <div class="span12">
      <div id="footer">
      </div>
    </div>
  </div>
</div>
</body>
```

创建 app.js 同时把 Backbone.js 的视图加入。

- ❑ headerView：菜单和应用一般信息。
- ❑ footerView：版权和联系信息。
- ❑ homeView：主页内容。

使用 Require.js 语法和 shim 插件加载 HTML 模板：

```
require([
'libs/text!header.html',
'libs/text!home.html',
'libs/text!footer.html'], function (
 headerTpl,
 homeTpl,
 footerTpl) {
```

应用的路由程序只有一个首页路由：

```
var ApplicationRouter = Backbone.Router.extend({
  routes: {
    "": "home",
    "*actions": "home"
  },
```

在开始做别的事情之前，把视图初始化一下，以便在整个应用里使用：

```
initialize: function () {
  this.headerView = new HeaderView();
  this.headerView.render();
  this.footerView = new FooterView();
  this.footerView.render();
},
```

首页路由的处理代码：

```
  home: function () {
    this.homeView = new HomeView();
    this.homeView.render();
  }
});
```

头部视图会添加到#header 元素，并且使用模板 headerTpl：

```
HeaderView = Backbone.View.extend({
  el: "#header",
  templateFileName: "header.html",
  template: headerTpl,
  initialize: function () {
  },
  render: function () {
    console.log(this.template)
    $(this.el).html(_.template(this.template));
  }
});
```

使用 jQuery.html 函数渲染 HTML：

```
FooterView = Backbone.View.extend({
  el: "#footer",
  template: footerTpl,
  render: function () {
```

```
    this.$el.html(_.template(this.template));
  }
});
```

首页视图定义使用`#content` DOM 元素：

```
HomeView = Backbone.View.extend({
  el: "#content",
  // template: "home.html"
  template: homeTpl,
  initialize: function () {
  },
  render: function () {
    $(this.el).html(_.template(this.template));
  }
});
```

创建一个新的实例，调用`Backbone.history.start()`启动应用：

```
  app = new ApplicationRouter();
  Backbone.history.start();
});
```

完整的 app.js 的代码：

```
require([
'libs/text!header.html',
//shim 插件使用示例
'libs/text!home.html',
'libs/text!footer.html'], function (
 headerTpl,
 homeTpl,
 footerTpl) {
  var ApplicationRouter = Backbone.Router.extend({
    routes: {
      "": "home",
      "*actions": "home"
    },
    initialize: function () {
      this.headerView = new HeaderView();
      this.headerView.render();
      this.footerView = new FooterView();
      this.footerView.render();
    },
    home: function () {
      this.homeView = new HomeView();
      this.homeView.render();
    }
  });
  HeaderView = Backbone.View.extend({
    el: "#header",
    templateFileName: "header.html",
    template: headerTpl,
    initialize: function () {
    },
```

5

```
    render: function () {
      console.log(this.template)
      $(this.el).html(_.template(this.template));
    }
  });

  FooterView = Backbone.View.extend({
    el: "#footer",
    template: footerTpl,
    render: function () {
      this.$el.html(_.template(this.template));
    }
  })
  HomeView = Backbone.View.extend({
    el: "#content",
    // template: "home.html"
    template: homeTpl,
    initialize: function () {
    },
    render: function () {
      $(this.el).html(_.template(this.template));
    }
  });
  app = new ApplicationRouter();
  Backbone.history.start();
});
```

上面是代码展示模板。所有的视图和路由程序都包含在内，提前加载模板，确保我们开始渲染的时候它们已经准备好了。

home.html 看起来是这样的：

- 表格展示的消息；
- Underscore.js 输出表格的行；
- 新消息表单。

我们使用 Twitter Bootstrap 库的结构（带有响应式组件），添加 `row-fluid` 和 `span12` 类：

```
<div class="row-fluid" id="message-board">
<div class="span12">
  <table class="table table-bordered table-striped">
    <caption>Chat</caption>
    <thead>
      <tr>
        <th class="span2">Username</th>
        <th>Message</th>
      </tr>
    </thead>
    <tbody>
```

下面简单介绍一下 Underscore.js 的模板，它把 JS 代码包裹在`<%`和`%>`标记里。`_.each` 是一个 Underscore.js 库（underscorejs.org/#each[1]）里的迭代函数，就像它的名字展现的那样，它遍历

① http://underscorejs.org/#each

数组和对象。

```
        <% if (models.length>0) {
          _.each(models, function (value,key, list) { %>
            <tr>
              <td><%= value.attributes.username %></td>
              <td><%= value.attributes.message %></td>
            </tr>
          <% });
        }
        else { %>
        <tr>
          <td colspan="2">No messages yet</td>
        </tr>
        <%}%>
      </tbody>
    </table>
  </div>
  </div>
```

新消息表单，我们使用 row-fluid 类，然后添加 input 元素：

```
<div class="row-fluid" id="new-message">
  <div class="span12">
    <form class="well form-inline">
      <input type="text"
        name="username"
        class="input-small"
        placeholder="Username"/>
      <input type="text" name="message"
        class="input-small"
        placeholder="Message Text"/>
      <a id="send" class="btn btn-primary">SEND</a>
    </form>
  </div>
</div>
```

完整的 home.html 模板代码：

```
<div class="row-fluid" id="message-board">
<div class="span12">
  <table class="table table-bordered table-striped">
    <caption>Chat</caption>
    <thead>
      <tr>
        <th class="span2">Username</th>
        <th>Message</th>
      </tr>
    </thead>
    <tbody>
      <% if (models.length>0) {
        _.each(models, function (value,key, list) { %>
          <tr>
            <td><%= value.attributes.username %></td>
            <td><%= value.attributes.message %></td>
```

```
      </tr>
    <% });
    }
    else { %>
    <tr>
      <td colspan="2">No messages yet</td>
    </tr>
    <%}%>
  </tbody>
</table>
</div>
</div>
<div class="row-fluid" id="new-message">
  <div class="span12">
    <form class="well form-inline">
      <input type="text"
        name="username"
        class="input-small"
        placeholder="Username"/>
      <input type="text" name="message"
        class="input-small"
        placeholder="Message Text"/>
      <a id="send" class="btn btn-primary">SEND</a>
    </form>
  </div>
</div>
```

现在来添加下面的组件：

- Parse.com 集合；
- Parse.com 模型；
- 发送或添加消息事件；
- 获取/展示消息函数。

可以通过使用 Parse.com SDK 强制添加一个 `className` 属性来适应 Backbone 兼容模型对象/类。（这个属性是通过 Parse.comWeb 接口的 Data Browser 看到的集合名。）

```
Message = Parse.Object.extend({
        className: "MessageBoard"
});
```

Backbone 集合与 Parse.com JavaScript SDK 对象兼容，使它指向模型：

```
MessageBoard = Parse.Collection.extend ({
        model: Message
});
```

首页视图需要有一个监听器监听“SEND”按钮的点击：

```
HomeView = Backbone.View.extend({
        el: "#content",
        template: homeTpl,
        events: {
                "click #send": "saveMessage"
        },
```

当创建 homeView 视图的时候，同时创建一个集合并且给它附加事件监听器：

```
initialize: function () {
        this.collection = new MessageBoard();
        this.collection.bind("all", this.render, this);
        this.collection.fetch();
        this.collection.on("add", function (message) {
                message.save(null, {
                        success: function (message) {
                                console.log('saved ' + message);
                        },
                        error: function (message) {
                                console.log('error');
                        }
                });
                console.log('saved' + message);
        })
},
```

“SEND”按钮被点击时调用 saveMessage()：

```
  saveMessage: function () {
          var newMessageForm = $("#new-message");
          var username = newMessageForm.
find('[name="username"]').
attr('value');
          var message = newMessageForm.find('[name="message"]').
find('[name="message"]').
attr('value');
          this.collection.add({
                  "username": username,
                  "message": message
                  });
  },
  render: function () {
          console.log(this.collection);
          $(this.el).html(_.template(
          this.template,
          this.collection
        ));
            }
```

最终我们修改的 app.js 是这样的：

```
/*
 这是一本关于 JavaScript 和 Node.js 的书，
 它将教你如何快速创建移动和 Web 应用，
 更多内容请访问：http://rapidprototypingwithjs.com
 */

require([
  'libs/text!header.html',
  'libs/text!home.html',
  'libs/text!footer.html'],
  function (
    headerTpl,
```

5

```
    homeTpl,
    footerTpl) {
  Parse.initialize(
    "your-parse-app-id",
    "your-parse-js-sdk-key");
  var ApplicationRouter = Backbone.Router.extend({
    routes: {
      "": "home",
      "*actions": "home"
    },
    initialize: function () {
      this.headerView = new HeaderView();
      this.headerView.render();
      this.footerView = new FooterView();
      this.footerView.render();
    },
    home: function () {
      this.homeView = new HomeView();
      this.homeView.render();
    }
  });

  HeaderView = Backbone.View.extend({
    el: "#header",
    templateFileName: "header.html",
    template: headerTpl,
    initialize: function () {
    },
    render: function () {
      $(this.el).html(_.template(this.template));
    }
  });

  FooterView = Backbone.View.extend({
    el: "#footer",
    template: footerTpl,
    render: function () {
      this.$el.html(_.template(this.template));
    }
  });
  Message = Parse.Object.extend({
    className: "MessageBoard"
  });
  MessageBoard = Parse.Collection.extend({
    model: Message
  });

  HomeView = Backbone.View.extend({
    el: "#content",
    template: homeTpl,
    events: {
      "click #send": "saveMessage"
    },
```

```
    initialize: function () {
      this.collection = new MessageBoard();
      this.collection.bind("all", this.render, this);
      this.collection.fetch();
      this.collection.on("add", function (message) {
        message.save(null, {
          success: function (message) {
            console.log('saved ' + message);
          },
          error: function (message) {
            console.log('error');
          }
        });
        console.log('saved' + message);
      })
    },
    saveMessage: function () {
      var newMessageForm = $("#new-message");
      var username =
        newMessageForm
          .find('[name="username"]')
          .attr('value');
      var message = newMessageForm
        .find('[name="message"]')
        .attr('value');
      this.collection.add({
        "username": username,
        "message": message
        });
    },
    render: function () {
      console.log(this.collection)
      $(this.el).html(_.template(
        this.template,
        this.collection
      ));
    }
  });

  app = new ApplicationRouter();
  Backbone.history.start();
});
```

完整的 Backbone.js 和 Parse.com Chat 应用源码在 rpjs/sdk[1]可以获取。

5.2 部署 Chat 到 PaaS

一旦确认了你的前端应用在本地工作正常了（使用或不使用本地 HTTP 服务器，如 MAMP 或 XAMPP），请将它部署到 Windows Azure 或者 Heroku。详细的部署说明可以参考第 3 章。

① https://github.com/azat-co/rpjs/tree/master/sdk

5.3　增强 Chat 应用

在上面的两个示例中，Chat 的功能非常基础。你可以通过增加更多特性来增强这个应用。

中级开发者可以添加的一些功能：

- 在展示之前通过 `updateAt` 属性对消息列表排序；
- 添加一个“Refresh”按钮来刷新消息列表；
- 在第一条消息之后在运行时内存或者一个会话里存储用户名；
- 给每一个消息添加一个“顶”的功能，并存储；
- 给每一个消息添加一个“踩”的功能，并存储。

高级开发者可以添加的一些功能：

- 添加一个用户集合；
- 防止同一个用户多次“顶”；
- 使用 Parse.com 的功能添加用户注册和登录动作；
- 在每一个当前用户创建的消息旁添加一个删除按钮“Delete Message”；
- 在每一个当前用户创建的消息旁添加一个编辑按钮“Edit Message”。

Part 3 第三部分

后端原型构建

本部分内容

第 6 章

Node.js 和 MongoDB

提要：展示 Node.js 的 Hello World 程序、Node.js 的一些重要的核心模块、NPM 工作流，以及在 Herohu 和 Windows Azure 上部署 Nodes.js 应用的详细命令；学习 MongoDB 及其 shell、运行时和数据库 Chat 应用；研究一个测试驱动开发的例子。

> 人人都可以写出计算机理解的代码。优秀的程序员写的是人可以理解的代码。
>
> ——马丁 · 福勒[1]

6.1 Node.js

6.1.1 创建Node.js的Hello World程序

首先需要检查一下你的电脑里是否装了 Node.js，在终端里输入下面的命令并运行：

```
$ node -v
```

写作本书时，最新的版本是 0.8.1。如果你没有安装 Node.js，或者你的版本比较落后，可以在 nodejs.org/#download[2]下载最新的版本。

按照惯例，你可以从 rpjs/hello[3]复制示例代码或者自己动手写。如果你想用后一种方式，首先创建一个名为 hello 的文件夹，然后创建名为 server.js 的文件，接下来一行一行地输入下面的代码。

这行代码会为服务器载入核心的 http 模块（稍后会有关于此模块的详细介绍）：

```
var http = require('http');
```

我们需要定义我们的 Node.js 服务器使用的端口。首先尝试从环境变量中获取，如果环境变

① http://en.wikipedia.org/wiki/Martin_Fowler

② http://nodejs.org/#download/

③ https://github.com/azat-co/rpjs/tree/master/hello

量中没有，则设置一个默认的值，代码如下：

```
var port = process.env.PORT || 1337;
```

创建一个服务器程序，它的回调函数包含了处理响应的代码：

```
var server = http.createServer(function (req, res) {
```

设置正确的首部和响应状态码：

```
res.writeHead(200, {'Content-Type': 'text/plain'});
```

输出“Hello World”和换行符号：

```
  res.end('Hello World\n');
});
```

设置服务监听的端口，并且在终端输出服务器地址及端口号：

```
server.listen(port, function() {
  console.log('Server is running at %s:%s ',
    server.address().address, server.address().port);
});
```

在终端里 server.js 所在的文件夹，运行下面的命令：

```
$ node server.js
```

用浏览器打开 localhost:1337[①]或 127.0.0.1:1337[②]，或者你在终端里看到的由 `console.log()` 函数输出的其他地址，你将在浏览器里看到“Hello World”。可以通过按下 Ctrl + C 快捷键关闭服务器。

注意

主文件的名字也有可能不是 server.js，比如有的是 index.js，有的是 app.js。如果你需要运行 app.js，使用`$ node app.js` 即可。

6.1.2 Node.js核心模块

与别的编程语言不同的是，Node.js 并没有自带一个庞大的标准库。Node.js 的核心模块是非常小的，其他的模块可以通过 NPM（Nodejs Package Manager）获取。主要的核心类、模块、方法和事件包含：

- http[③]

---

① http://localhost:1337

② http://127.0.0.1:1337

③ http://nodejs.org/api/http.html#http_http

- util[①]
- querystring[②]
- url[③]
- fs[④]

http[⑤]

这个模块主要负责 Node.js HTTP 服务器。下面是它的主要方法。

- `http.createServer()`：返回一个新的 Web 服务器对象。
- `http.listen()`：开始在特定的端口和主机名接收连接。
- `http.createClient()`：node 应用可以作为客户端并且向别的服务端发送请求。
- `http.ServerRequest()`：收到的请求会传递给如下请求处理函数。
  - `data`：收到信息主体时触发的事件。
  - `end`：每个请求结束时只触发一次的事件。
  - `request.method()`：字符串作为请求的方法名。
  - `request.url()`：请求的 URL 字符串。
- `http.ServerResponse()`：HTTP 服务器内部创建的对象，而不是由用户创建的，作为请求处理函数的输出。
  - `response.writeHead()`：向请求发出一个响应首部。
  - `response.write()`：给请求发送响应头。
  - `response.end()`：发出并结束响应体。

util[⑥]

提供用来调试的工具函数，例如下面这个函数。

- `util.inspect()`：返回一个对象的字符串表示，这在调试的时候很有用。

querystring[⑦]

提供对查询字符串进行处理的工具函数，它包含下面这些方法。

---

① http://nodejs.org/api/util.html
② http://nodejs.org/api/querystring.html
③ http://nodejs.org/api/url.html
④ http://nodejs.org/api/fs.html
⑤ http://nodejs.org/api/http.html#http_http
⑥ http://nodejs.org/api/util.html
⑦ http://nodejs.org/api/querystring.html

- querystring.stringify()：把一个对象序列化成查询字符串。
- querystring.parse()：把一个查询字符串反序列化成对象。

url[①]

包含用于 URL 处理和解析的工具函数，例如下面这个函数。

- parse()：处理一个 URL 字符串，并返回一个对象。

fs[②]

用于处理文件系统操作，比如读写文件。这个库里既有同步的函数也有异步的函数。下面是它包含的一些方法。

- fs.readFile()：异步读取一个文件。
- fs.writeFile()：将数据异步写入一个文件。

核心模块不用安装或者下载。如果想在你的程序中使用它们，只需要用下面的代码：

```
var http = require('http');
```

非核心的模块可以用下列方法找到。

- npmjs.org[③]：Node Package Manager 注册库。
- GitHub hosted list[④]：Joyent 维护的 Node.js 模块列表。
- nodetoolbox.com[⑤]：基于统计数据的注册库。
- Nipster[⑥]：NPM 搜索工具。
- Node Tracking[⑦]：基于 GitHub 统计数据的注册库。

如果你想了解如何写一个属于自己的模块，可以看看这篇文章：Your first Node.js module[⑧]。

6.1.3 NPM

NPM 可以为你管理模块依赖和安装模块。Node.js 的安装程序默认带了 NPM，它的网址是 npmjs.org[⑨]。

① http://nodejs.org/api/url.html
② http://nodejs.org/api/fs.html
③ https://npmjs.org
④ https://github.com/joyent/node/wiki/Modules
⑤ http://nodetoolbox.com/
⑥ http://eirikb.github.com/nipster/
⑦ http://nodejsmodules.org
⑧ http://cnnr.me/blog/2012/05/27/your-first-node-dot-js-module/
⑨ https://npmjs.org

package.json 包含了我们的 Node.js 程序的元信息，比如版本号、作者，最重要的是我们的应用程序的依赖。所有这些信息以 JSON 对象的格式保存，NPM 会读取它。

如果你想安装在 package.json 里定义的包和依赖，输入：

```
$ npm install
```

一般 package.json 文件如下所示：

```
{
   "name": "Blerg",
   "description": "Blerg blerg blerg.",
   "version": "0.0.1",
   "author": {
      "name" : "John Doe",
      "email" : "john.doe@gmail.com"
   },
   "repository": {
     "type": "git",
     "url": "http://github.com/johndoe/blerg.git"
   },
   "engines": [
     "node >= 0.6.2"
   ],
   "license" : "MIT",
   "dependencies": {
     "express": ">= 2.5.6",
     "mustache": "0.4.0",
     "commander": "0.5.2"
   },
   "bin" : {
     "blerg" : "./cli.js"
   }
}
```

把一个包更新到最新版本，或者由 package.json 中的版本说明所定义的最新版本，请使用：

```
$ npm update name-of-the-package
```

或者，单独安装一个模块：

```
$ npm install name-of-the-package
```

本书示例中用到的唯一不属于 Node.js 核心模块的模块是 mongodb，稍后的章节将介绍如何安装它。

Heroku 在服务器上运行 NPM 需要 package.json。

更多关于 NPM 的信息，请参阅这篇文章：Tour of NPM[①]。

---

① http://tobyho.com/2012/02/09/tour-of-npm/

6.1.4 部署Hello World到PaaS

Heroku 和 Windows Azure 的部署都需要有一个 Git 仓库。在你的项目的根目录创建一个 Git 仓库，需输入并运行下面的命令：

```
$ git init
```

Git 将会创建一个隐藏的.git 文件夹。现在我们来添加文件，然后做第一次提交：

```
$ git add .
$ git commit -am "first commit"
```

技巧

如果想在 Mac OS X Finder 应用里查看隐藏文件，可以在终端窗口执行命令：`defaults write com.apple.finder AppleShowAllFiles -bool true`。恢复隐藏文件的命令是：`defaults write com.apple.finder AppleShowAllFiles -bool false`。

6.1.5 部署到Windows Azure

为了把 Hello World 程序部署到 Windows Azure，必须添加一个 Git 远程地址。你可以从 Windows Azure Portal 上的 Web Site 下复制地址，然后使用下面的命令：

```
$ git remote add azure yourURL
```

使用下面的命令来推送你的代码：

```
$ git push azure master
```

如果一切正常，你可以在命令行里看到成功的日志，用浏览器查看你的 Windows Azure Web Site URL 能看到"Hello World"。

推送本地的一些修改，只需要执行：

```
$ git add .
$ git commit -m "changing to hello azure"
$ git push azure master
```

更详细的指导可以查看这篇教程：Build and deploy a Node.js web site to Windows Azure[①]。

6.1.6 部署到Heroku

为了部署到 Heroku，我们还需要再创建两个文件：Procfile 和 package.json。你可以从 rpjs/hello[②]

① http://www.windowsazure.com/en-us/develop/nodejs/tutorials/create-a-website-(mac)/

② https://github.com/azat-co/rpjs/tree/master/hello

获得源码，或者自己写一份。

Hello World 程序的目录结构如下所示：

```
/hello
  -package.json
  -Procfile
  -server.js
```

Procfile 用来描述在 Heroku 平台上运行你的项目里的哪一个命令。事实上，它告诉 Heroku 该运行什么。在这个例子里 Procfile 只有一行：

```
web: node server.js
```

这个例子里，我们使用了一个很简单的 package.json：

```
{
  "name": "node-example",
  "version": "0.0.1",
  "dependencies": {
  },
  "engines": {
    "node": ">=0.6.x"
  }
}
```

所有的文件都在项目文件夹里之后，我们使用 Git 来部署应用程序。命令和 Windows Azure 的命令很类似，只是需要添加 Git 远程库地址，还需要使用下面的命令创建 Cedar stack：

```
$ heroku create
```

执行完毕后，我们可以提交并且推送到远程库了：

```
$ git push heroku master
$ git add .
$ git commit -am "changes"
$ git push heroku master
```

如果一切正常，可以在终端里可以看到成功的日志，并且通过浏览器访问你的 Heroku app 网址，可以看到“Hello World”。

6.2　Chat：运行时内存版本

为了遵循 KISS[1]（Keep It Simple, Stupid）编程原则，第一个版本的 Chat 服务端程序仅把信息保存在运行时内存中。这意味着，每一次我们启动或者重置服务器的时候，数据都会清空。

---

① http://en.wikipedia.org/wiki/KISS_principle

我们从一个简单的测试用例开始，来说明测试驱动的开发方式。全部的代码可以在 rpjs/test[①] 文件夹下找到。

6.3 Chat 的测试用例

我们有如下两个方法：

(1) `getMessages()` 方法用来处理 GET `/message` 请求，它以 JSON 数组对象的格式返回所有的消息；

(2) `addMessage()` 方法用来处理 POST `/messages` 请求，它会添加一个包含 `name` 和 `message` 属性的信息。

先创建一个空的 mb-server.js 文件。接下来让我们创建一个包含下面内容的测试文件 test.js：

```
var http = require('http');
var assert = require('assert');
var querystring = require('querystring');
var util = require('util');

var messageBoard = require('./mb-server');

assert.deepEqual('[{"name":"John","message":"hi"}]',
  messageBoard.getMessages());
assert.deepEqual ('{"name":"Jake","message":"gogo"}',
  messageBoard.addMessage ("name=Jake&message=gogo"));
assert.deepEqual('[{"name":"John","message":"hi"},{"name":"Jake","message":"gogo"}]',
  messageBoard.getMessages("name=Jake&message=gogo"));
```

请一定牢记，这里比较的仅仅是简单的字符串，而不是 JavaScript 对象！所以，空格、引号和大小写是有区别的。你也可以尝试像这样把字符串解析成 JSON 对象，进行更“智能”的比较：

```
JSON.parse(str);
```

为了验证我们的假设，我们使用 Node.js 核心模块 assert[②]。它提供了一些非常有用的方法，比如 `equal()`、`deepEqual()` 等。

以下高级库可以提供 TDD 或者 BDD 方式的测试，比如：

- Should[③]
- Expect[④]

要应用更多测试驱动开发和自动化测试，可以使用下面的库和模块：

① https://github.com/azat-co/rpjs/tree/master/test

② http://nodejs.org/api/assert.html

③ https://github.com/visionmedia/should.js/

④ https://github.com/LearnBoost/expect.js/

- ❑ Mocha[①]
- ❑ NodeUnit[②]
- ❑ Jasmine[③]
- ❑ Vows[④]
- ❑ Chai[⑤]

现在可以把Hello World的代码复制到mb-server.js里或者保持它的内容是空的。若在终端里执行test.js：

```
$ node test.js
```

会有错误出现，大概会像这样：

```
TypeError: Object #<Object> has no method 'getMessages'
```

这是正常的，因为我们还没有写getMessages()方法。接下来，我们通过添加两个方法来让程序达到我们的期望：一个方法获取消息列表，另一个方法把消息添加到集合里。

mb-server.js的全局exports对象：

```
exports.getMessages = function() {
  return JSON.stringify(messages);
};
exports.addMessage = function (data){
  messages.push(querystring.parse(data));
  return JSON.stringify(querystring.parse(data));
};
```

到目前为止，还没有太神奇的代码吧？我们使用一个数组来保存消息列表：

```
var messages=[];
//这个数组将保存消息
messages.push({
  "name":"John",
  "message":"hi"
});
//用于测试输出列表的简单信息
```

提示

一般情况下，像模拟数据这样的代码应该放在test/spec文件里，而不是直接放到主程序里。

---

① http://visionmedia.github.com/mocha/

② https://github.com/caolan/nodeunit

③ http://pivotal.github.com/jasmine/

④ http://vowsjs.org/

⑤ http://chaijs.com/

我们的服务器代码看上去有一点意思了。为了获取消息列表，根据 REST 方法论，我们需要发起一个 GET 请求。创建/添加一条消息，应该是一个 POST 请求。所以在我们的 createServer 对象里，我们应该添加 req.method() 和 req.url() 检查 HTTP 请求的类型和 URL 路径。

加载 http 模块：

```
var http = require('http');
```

使用 util 和 querystring 模块里的一些辅助函数（用来序列化和反序列化对象及查询字符串）：

```
var util = require('util');
var querystring = require('querystring');
```

创建一个服务器并且暴露到模块外，这样就可以供别的代码使用（即 test.js）：

```
exports.server=http.createServer(function (req, res) {
```

在请求的回调函数里，我们需要检查请求的方法是否是 POST，URL 是否是 messages/create.json：

```
if (req.method == "POST" &&
  req.url == "/messages/create.json") {
```

如果上面的条件是 true，我们添加一个消息到数组里。然而，数据在添加前必须被转换为字符串类型（编码是 UTF-8），因为它是一个 Buffer 类型：

```
var message = '';
req.on('data', function(data, message){
  console.log(data.toString('utf-8'));
  message = exports.addMessage(data.toString('utf-8'));
```

输出日志帮助我们在终端里查看服务器的状态：

```
})
console.log(util.inspect(message, true, null));
console.log(util.inspect(messages, true, null));
```

请求的响应应该是文本格式，同时状态码是 200（很好）：

```
res.writeHead(200, {
  'Content-Type': 'text/plain'
 });
```

输出新创建的消息：

```
  res.end(message);
}
```

如果请求的方法是 GET 并且 URL 是/messages/list.json，那就输出消息列表：

```
if (req.method == "GET" &&
  req.url == "/messages/list.json") {
```

获取消息列表：

6

```
var body = exports.getMessages();
```

响应体将会包含我们的输出：

```
  res.writeHead(200, {
    'Content-Length': body.length,
    'Content-Type': 'text/plain'
  });
  res.end(body);
}
else {
```

设置正确的首部和状态码：

```
res.writeHead(200, {
  'Content-Type': 'text/plain'
});
```

如果不符合上面两种情况，则输出一个带换行符的字符串：

```
    res.end('Hello World\n');
  };
  console.log(req.method);

}).listen(1337, "127.0.0.1");
```

在终端里设置我们的服务器端口和 IP 地址：

```
console.log('Server running at http://127.0.0.1:1337/');
```

通过 exports 暴露出方法，这样就可以在单元测试的 test.js（导出关键字）里使用，这个方法会返回字符串/文本格式的消息数组：

```
exports.getMessages = function() {
  return JSON.stringify(messages);
};
```

addMessage() 使用 querystring 的解析和反序列化方法把一个字符串转换为一个 JavaScript 对象：

```
exports.addMessage = function (data){
  messages.push(querystring.parse(data));
```

同时返回一个字符串格式的消息：

```
  return JSON.stringify(querystring.parse(data));
};
```

下面是 mb-server.js 的完整代码，在 rpjs/test[1]里也可以找到它：

```
var http = require('http');
//加载 http 模块
var util = require('util');
```

[1] https://github.com/azat-co/rpjs/tree/master/test

```
//有用的函数
var querystring = require('querystring');
//加载 querystring 模块,
//需要用它来序列化和反序列化对象及请求字符串

var messages = [];
//这个数组将保存消息
messages.push({
  "name": "John",
  "message": "hi"
});
//用于测试输出列表的简单信息

exports.server = http.createServer(function(req, res) {
//创建服务器
  if (req.method == "POST" &&
    req.url == "/messages/create.json") {
    //如果请求方法是 POST，并且 URL 是 messages/,
    //添加这个消息到数组里
    var message = '';
    req.on('data', function(data, message) {
      console.log(data.toString('utf-8'));
      message = exports.addMessage(data.toString('utf-8'));
      //data 是 Buffer 类型并且必须使用 UTF-8 编码转为字符串
      //添加到消息数组
    })
    console.log(util.inspect(message, true, null));
    console.log(util.inspect(messages, true, null));
    //调试信息输出到终端
    res.writeHead(200, {
      'Content-Type': 'text/plain'
    });
    //设置正确的响应头及状态码
    res.end(message);
    //输出信息，应该添加对象 id
  }
  if (req.method == "GET" &&
    req.url == "/messages/list.json") {
  //如果请求方法是 GET，并且 URL 是/messages
  //输出消息列表
    var body = exports.getMessages();
    //body 变量保存输出
    res.writeHead(200, {
      'Content-Length': body.length,
      'Content-Type': 'text/plain'
    });
    res.end(body);
  }
  else {
    res.writeHead(200, {
      'Content-Type': 'text/plain'
    });
    //设置正确的响应头及状态码
    res.end('Hello World\n');
```

```
  };
  console.log(req.method);
  //输出字符串同时附加换行符
}).listen(1337, "127.0.0.1");
//设置服务器的端口和IP地址
console.log('Server running at http://127.0.0.1:1337/');

exports.getMessages = function() {
  return JSON.stringify(messages);
  //以字符串文本的形式输出消息
};
exports.addMessage = function(data) {
  messages.push(querystring.parse(data));
  //用解析和反序列化器把一个请求串转成JavaScript对象
  return JSON.stringify(querystring.parse(data));
  //以JSON字符串的形式输出新消息
};
```

用浏览器打开 localhost:1337/messages/list.json[①]检查一下，你会看到一条测试用的消息。另外，也可以在终端里通过 `curl` 命令测试：

```
$ curl http://127.0.0.1:1337/messages/list.json
```

使用命令行接口模拟一次 `POST` 请求：

```
curl -d "name=BOB&message=test" http://127.0.0.1:1337/messages/create.json
```

你将会在服务器终端窗口看到输出；然后刷新 localhost:1337/messages/list.json[②]，会看到一条"test"新消息。毫无疑问，三个测试用例都应该通过了。

随着方法、要解析的路由和条件越来越多，程序也会变得越来越大。这时使用框架的好处就显现出来了。它们提供处理请求的辅助函数和其他有用的东西，比如 session、静态文件支持等。在这个例子里，我们特意没有使用诸如 Express（http://expressjs.com/）、Restify（http://mcavage.github.com/node-restify/）这样的框架。下面是一些值得关注的 Node.js 框架。

- Derby[③]：MVC框架，用来构建在Node.js和浏览器里运行的实时、协作应用。
- Express.js[④]：最健壮、测试最完善、使用最多的Node.js框架。
- Restify[⑤]：构建REST API服务器的轻量级框架。
- Sails.js[⑥]：MVC Node.js框架。
- hapi[⑦]：在Express.js基础上构建的Node.js框架。

---

① http://localhost:1337/messages/list.json

② http://localhost:1337/messages/list.json

③ http://derbyjs.com

④ http://expressjs.com

⑤ http://mcavage.github.com/node-restify/

⑥ http://sailsjs.com

⑦ http://spumko.github.io

- Connect[1]：node 中间件框架，内置超过 18 个中间件，并且拥有大量可选的第三方中间件。
- GeddyJS[2]：一个简单的结构化的 Node MVC Web 框架。
- CompoundJS[3]（ex-RailswayJS）：在 ExpressJS 基础上构建的 Node.js MVC 框架。
- Tower.js[4]：为 Node.js 和浏览器开发的全栈 Web 框架。
- Meteor[5]：只需用很少的时间就可以构建出高质量 Web 应用的开源平台。

我们的程序可以改进的地方有：

- 改进已有的测试用例，对比对象而不是字符串；
- 把测试数据从 mb-server.js 里移到 test.js 里；
- 给前端代码添加测试用例，比如投票、用户登录；
- 添加方法，完善前端代码，比如投票、用户登录；
- 为每一个消息生成唯一的 ID，用散列而不是数组存储消息；
- 安装 Mocha，并且使用它重构 test.js。

到目前我们还是把消息都存放在了程序内存里，所以每一次程序重启，消息就丢了。为了解决这个问题，我们需要有一个永久存储，其中一种方式就是使用数据库，比如 MongoDB。

6.4 MongoDB

6.4.1 MongoDB Shell

如果你的环境还没有准备好，请先到 mongodb.org/downloads[6]下载安装最新版的 MongoDB。更多的内容可以参考 2.1.6 节。

现在让我们从你解压的文件夹的位置开始，启动 `mongod` 服务：

```
$ ./bin/mongod
```

可以在终端里看到一些信息，也可以通过访问 localhost:28017[7]看到信息。

同样，在刚才解压的文件夹中，另开一个新窗口（非常重要！），使用 MongoDB 的命令行或 `mongo`，运行命令：

---

① http://www.senchalabs.org/connect/
② http://geddyjs.corg
③ http://compoundjs.com
④ http://towerjs.org
⑤ http://meteor.com
⑥ http://www.mongodb.org/downloads
⑦ http://localhost:28017

```
$ ./bin/mongo
```

你会看到像下面一样的信息，这和你安装的 MongoDB 命令行的版本有关：

```
MongoDB shell version: 2.0.6
connecting to: test
```

测试一下数据库，它的接口很像 JavaScript，我们使用命令 `save` 和 `find` 来进行保存和查找：

```
> db.test.save( { a: 1 } )
> db.test.find()
```

更多详细指导，请查看 2.1.6 节的内容。

下面是一些有用的 MongoDB shell 命令：

```
> help
> show dbs
> use board
> show collections
> db.messages.remove();
> var a = db.messages.findOne();
> printjson(a);
> a.message = "hi";
> db.messages.save(a);
> db.messages.find({});
> db.messages.update({name: "John"},{$set: {message: "bye"}});
> db.messages.find({name: "John"});
> db.messages.remove({name: "John"});
```

完整的 MongoDB 交互式 shell 可以在 mongodb.org 中找到：Overview - The MongoDB Interactive Shell①。

6.4.2 MongoDB原生驱动

我们使用 MongoDB 的原生 Node.js 驱动（https://github.com/christkv/node-mongodb-native）来访问 MongoDB。完整的文档可以在 http://mongodb.github.com/node-mongodb-native/apigenerated/db.html 中看到。

首先来安装 MongoDB 的原生 Node.js 驱动，使用：

```
$ npm install mongodb
```

更多详情请浏览：http://www.mongodb.org/display/DOCS/node.JS。

不要忘记把这个依赖添加到 package.json 里②：

---

① http://www.mongodb.org/display/DOCS/Overview+-+The+MongoDB+Interactive+Shell
② 这里可以简单使用 `npm install -save mongodb` 自动把依赖添加到 package.json。——译者注

```
{
  "name": "node-example",
  "version": "0.0.1",
  "dependencies": {
       "mongodb":"",
       ...
  },
  "engines": {
    "node": ">=0.6.x"
  }
}
```

你自己开发时也可以使用一些别的封装好的模块，它们是原生驱动的扩展。

- Mongoskin[①]：node-mongodb-native 的未来包装层。
- Mongoose[②]：带有可选的建模功能的异步 JavaScript 驱动器。
- Mongolia[③]：轻量级 MongoDB ORM/驱动器包装器。
- Monk[④]：一个小型的包装层，给 Node.js 里使用 MongoDB 提供了简单且更易用的增强。

下面的小例子测试是否能通过 Node.js 脚本连上 MongoDB 实例。

安装完这个库后，我们可以在 db.js 文件里引入它：

```
var util = require('util');
var mongodb = require ('mongodb');
```

这是一种创建与 MongoDB 服务器连接的方式，Db 变量保存着一个特定的端口和主机地址的数据库连接引用：

```
var Db = mongodb.Db;
var Connection = mongodb.Connection;
var Server = mongodb.Server;
var host = '127.0.0.1';
var port = 27017;

var db=new Db ('test', new Server(host,port, {}));
```

打开一个链接：

```
db.open(function(e,c){
      //这里对数据库进行一些操作
  // console.log (util.inspect(db));
  console.log(db._state);
  db.close();
});
```

---

① https://github.com/guileen/node-mongoskin

② http://mongoosejs.com

③ https://github.com/masylum/mongolia

④ https://github.com/LearnBoost/monk

这段代码在 rpjs/db/db.js[1]文件夹中可以看到。如果你运行它，会在终端里输出“connected”。如果你对它有怀疑并且想检查一下这个对象的属性，可以使用 util 模块里的一个方法：

```
console.log(util.inspect(db));
```

6.4.3　MongoDB on Heroku：MongoHQ

在本地成功输出“connected”之后，是时候做些小改动，然后作为服务部署到平台上了，即 Heroku。

我们推荐使用 MongoHQ 扩展[2]，它是 MongoHQ[3]技术的一部分。它提供了一个浏览器界面来查找和修改数据及集合。更多的信息可以在这里查看：https://devcenter.heroku.com/articles/mongohq。

注意

尽管选择免费版本，使用 MongoHQ 的时候还是需要你提供信用卡信息。当然，它是不会向你收费的。

为了连接到数据库服务器，需要使用一个数据库连接 URL（也叫 MongoHQ URL/URI），它的目的是把连接需要的所有信息组织成一个字符串。

数据库连接字符串（MONGOHQ_URL）的格式如下：

```
mongodb://user:pass@server.mongohq.com/db_name
```

可以从 Heroku 的网站上复制 MongoHQ URL 字符串（然后硬编码），也可以从 Node.js 的环境变量对象获取：

```
process.env.MONGOHQ_URL;
```

或：

```
var connectionUri = url.parse(process.env.MONGOHQ_URL);
```

提示

可以通过全局对象 process 上的 env 变量来访问环境变量。这些变量一般用来传递数据库主机名、端口、密码、API 密钥、端口号和其他系统信息，这些信息不应该直接硬编码进主逻辑。

为了让我们的代码在本地和 Heroku 上同时运行，可以使用逻辑操作符“或”（||），如果环

① https://github.com/azat-co/rpjs/blob/master/db/db.js

② https://addons.heroku.com/mongohq

③ https://www.mongohq.com/home

境变量是 undefined，那么使用本地主机和端口，代码如下：

```
var port = process.env.PORT || 1337;
var dbConnUrl = process.env.MONGOHQ_URL ||
  'mongodb://@127.0.0.1:27017';
```

下面是更新的能跨环境的 db.js 文件：

```
var url = require('url');
var util = require('util');
var mongodb = require ('mongodb');
var Db = mongodb.Db;
var Connection = mongodb.Connection;
var Server = mongodb.Server;

var dbConnUrl = process.env.MONGOHQ_URL ||
'mongodb://127.0.0.1:27017';
var host = url.parse(dbConnUrl).hostname;
var port = new Number(url.parse(dbConnUrl).port);
var db=new Db ('test', new Server(host,port, {}));
db.open(function(e,c){
  // console.log (util.inspect(db));
  console.log(db._state);
  db.close();
});
```

通过添加 MONGOHQ_URL 对 db.js 进行改造，现在我们来初始化 Git 仓库、创建 Heroku 程序、添加 MongoHQ 扩展，然后使用 Git 部署应用程序。

利用和之前例子一样的步骤来创建一个 Git 仓库：

```
git init
git add .
git commit -am 'initial commit'
```

创建 Heroku 程序的 Cedar Stack：

```
$ heroku create
```

如果一切正常，你会看到一条带有 Heroku 程序名（和 URL）的消息和一条添加了远程的消息。本地 Git 仓库存在远程很重要，你可以通过下面的命令查看已经添加的远程列表：

```
git remote show
```

在已经存在的 Heroku 程序上安装免费的 MongoHQ，使用：

```
$ heroku addons:add mongohq:sandbox
```

如果你知道程序的名字，也可以登录 addons.heroku.com/mongohq[1]，为该 Heroku 程序选择免费的 MongoHQ。

① https://addons.heroku.com/mongohq

如果你的 db.js 和修改后的 db.js 正常工作了，让我们来添加一个 HTTP 服务器，在浏览器里展示“connected”消息，而不是在终端里。为了达到这个效果，我们需要在数据库连接的回调里封装一下 server 对象实例：

```
...
db.open(function(e, c) {
        // console.log (util.inspect(db));
        var server = http.createServer(function(req, res) {
    //创建服务器
          res.writeHead(200, {'Content-Type': 'text/plain'});
    //设置正确的响应头及状态码
          res.end(db._state);
    //输出字符串同时附加换行符
        });
        server.listen(port, function() {
                console.log('Server is running at %s:%s ',
    server.address().address,
    server.address().port);
    //设置服务器的端口与 IP 地址
        });
        db.close();
});
...
```

最终可以用 rpjs/db 的 app.js 来部署：[1]

```
/*
这是一本关于 JavaScript 和 Node.js 的书，
它将教你如何快速创建移动和 Web 应用，
更多内容请访问：http://rapidprototypingwithjs.com
*/
var util = require('util');
var url = require('url');
var http = require('http');
var mongodb = require('mongodb');
var Db = mongodb.Db;
var Connection = mongodb.Connection;
var Server = mongodb.Server;
var port = process.env.PORT || 1337;
var dbConnUrl = process.env.MONGOHQ_URL ||
  'mongodb://@127.0.0.1:27017';
var dbHost = url.parse(dbConnUrl).hostname;
var dbPort = new Number(url.parse(dbConnUrl).port);
console.log(dbHost + dbPort);
var db = new Db('test', new Server(dbHost, dbPort, {}));
db.open(function(e, c) {
  // console.log (util.inspect(db));
  // 创建服务器
  var server = http.createServer(function(req, res) {
    //设置正确的响应头及状态码
```

① https://github.com/azat-co/rpjs/blob/master/db

```
    res.writeHead(200, {
      'Content-Type': 'text/plain'
    });
    //输出字符串同时附加换行符
    res.end(db._state);
  });
  //设置服务器的端口与IP地址
  server.listen(port, function() {
    console.log(
      'Server is running at %s:%s ',
    server.address().address,
    server.address().port);
  });
  db.close();
});
```

部署完成后，打开 Herorku 提供的 URL，你可以看到消息“connected”。

下面是介绍如何用 Node.js 代码来使用 MongoDB 的手册：mongodb.github.com/node-mongodb-native/api-articles/nodekoarticle1.html[①]。

另一种方法是使用 MongoHQ 模块，可以从 github.com/MongoHQ/mongohq-nodejs[②]获取。

下面的例子演示了 mongodb 库的另一种用法，输出集合和文档数量。完整的源码从 rpjs/db/collections.js[③]获取：

```
var mongodb = require('mongodb');
var url = require('url');
var log = console.log;
var dbUri = process.env.MONGOHQ_URL ||'mongodb://localhost:27017/test';

var connectionUri = url.parse(dbUri);
var dbName = connectionUri.pathname.replace(/^\//, '');

mongodb.Db.connect(dbUri, function(error, client) {
  if (error) throw error;

  client.collectionNames(function(error, names) {
    if (error) throw error;

    //输出所有集合的名字
    log("Collections");
    log("===========");
    var lastCollection = null;
    names.forEach(function(colData) {
      var colName = colData.name.replace(dbName + ".", '');
      log(colName);
```

① http://mongodb.github.com/node-mongodb-native/api-articles/nodekoarticle1.html

② https://github.com/MongoHQ/mongohq-nodejs

③ https://github.com/azat-co/rpjs/blob/master/db/collections.js

```
      lastCollection = colName;
    });
    if (!lastCollection) return;
    var collection = new mongodb.Collection(client, lastCollection);
    log("\nDocuments in " + lastCollection);
    var documents = collection.find({}, {limit: 5});

    //输出找到的所有文档的数量
    documents.count(function(error, count) {
      log(" " + count + " documents(s) found");
      log("====================");

      //输出前五个文档
      documents.toArray(function(error, docs) {
        if (error) throw error;

        docs.forEach(function(doc) {
          log(doc);
        });

        //关闭连接
        client.close();
      });
    });
  });
});
```

通过 `var log = console.log;`来使用 `console.log()`的快捷方式，并且通过 `if(!lastCollection) return;`将 `return` 用作一个控制流。

6.4.4 BSON

Binary JSON，也叫 BSON，它是 MongoDB 使用的一种专有的数据类型。在格式上它有点像 JSON，但是支持更多复杂的数据类型。

警告

关于 BSON 需要注意的地方：MongoDB 里的 `ObjectId` 和 MongoDB 的原生 Node.js 驱动里的 `ObjectID` 是一样的。请确保你使用的是正确的那个，否则将会出错。更多相关类型参见：ObjectId in MongoDB[①] vs Data Types in MongoDB Native Node.js Drier[②]。关于 `mongodb.ObjectID()`的 Node.js 代码例子：`collection.findOne({_id: new ObjectID(idString)}, console.log) // ok`。另一方面，在 MongoDB 命令行里，我们使用：`db.messages.findOne({_id:ObjectId(idStr)});`。

① http://www.mongodb.org/display/DOCS/Object+IDs

② https://github.com/mongodb/node-mongodb-native/#data-types

6.5 Chat：MongoDB 版本

我们已经设置好了令 Node.js 程序同时在本地和 Heroku 运行所需要的东西。源码可以在这里找到：rpjs/mongo[1]。程序的结构很简单：

```
/mongo
  -web.js
  -Procfile
  -package.json
```

下面是 web.js 的代码，首先我们引入一些库：

```
var http = require('http');
var util = require('util');
var querystring = require('querystring');
var mongo = require('mongodb');
```

接下来使用一个神奇的字符串来连接 MongoDB：

```
var host = process.env.MONGOHQ_URL || "mongodb://@127.0.0.1:27017/twitter-clone";
//MONGOHQ_URL=mongodb://user:pass@server.mongohq.com/db_name
```

注意

URI/URL 格式里包含的数据库名是可选的，我们的集合会保存在那里。你完全可以按照自己的意愿修改它，比如改为 rpjs 或者 test。

我们把所有的逻辑以回调函数的形式放在一个打开的连接里：

```
mongo.Db.connect(host, function(error, client) {
  if (error) throw error;
  var collection = new mongo.Collection(
    client,
    'messages');
  var app = http.createServer(function(request, response) {
    if (request.method === "GET" &&
      request.url === "/messages/list.json") {
      collection.
        find().
        toArray(function(error, results) {
        response.writeHead(200, {
          'Content-Type': 'text/plain'
        });
        console.dir(results);
        response.end(JSON.stringify(results));
      });
    };
    if (request.method === "POST" &&
      request.url === "/messages/create.json") {
      request.on('data', function(data) {
        collection.insert(
```

① https://github.com/azat-co/rpjs/tree/master/mongo

```
            querystring.parse(data.toString('utf-8')),
            {safe: true},
            function(error, obj) {
              if (error) throw error;
              response.end(JSON.stringify(obj));
            }
          )
        })
      };
    });
    var port = process.env.PORT || 5000;
    app.listen(port);
  })
```

注意

集合/实体名后面的那些单词不是必需的，即不用/messages/list.json 和 /messages/create.json，我们可以只使用/messages 来处理所有的 HTTP 方法，如 GET、POST、PUT 以及 DELETE。如果你这样修改了，记得同时修改你的 CURL 命令和前端代码。

通过下面的 CURL 终端命令来测试：

```
curl http://localhost:5000/messages/list.json
```

或者用浏览器打开 http://locahost:5000/messages/list.json。

它会返回一个空数组：[]，这是正常的。接下来 POST 产生一个新消息：

```
curl -d "username=BOB&message=test" http://localhost:5000/messages/create.json
```

现在肯定可以看到一个响应，其中包含新创建元素的 ObjectID，例如：[{"username":"BOB","message":id":"51edcad45862430000000001"}]。你的 ObjectID 可能和这个不一样。

如果在本地一切工作正常，请尝试把它部署到 Heroku。

把 http://localhost/或者 http://127.0.0.1 替换为你的 Heroku 程序的 URL，就可以使用同样的 CURL[①]命令来测试在 Heroku 上的程序，如下所示：

```
$ curl http://your-app-name.herokuapp.com/messages/list.json
$ curl -d "username=BOB&message=test"
  http://your-app-name.herokuapp.com/messages/create.json
```

我们可以通过 Mongo shell 的$ mongo 终端命令，然后使用 twitter-clone 和 db.messages.find()，再一次检查一下数据库。另外，我们也可以使用 MongoHub[②]、mongoui[③]、mongo-express[④]，

① http://curl.haxx.se/docs/manpage.html

② https://github.com/bububa/MongoHub-Mac

③ https://github.com/azat-co/mongoui

④ https://github.com/andzdroid/mongo-express

或者使用 heroku.com 上的 MongoHQ Web 界面。

如果你想使用别的域名，而不是 http://your-app-name.herokuapp.com，需要做两件事情：

(1) 告诉 Heroku 你的域名：

```
$ heroku domains:add www.your-domain-name.com
```

(2) 在你的 DNS 管理工具里添加一条 CNAME DNS 记录，并且指向 http://your-app-name.herokuapp.com。

更多关于自定义域名的信息可以在这里找到：devcenter.heroku.com/articles/custom-domains[①]。

提示

为了更高产和更高效地开发，我们需要尽可能的自动化，即使用测试而不是 CURL 命令。在第 8 章有一篇文章是关于如何使用 Mocha 库的文章，该库和 superagent 或者 request 库一起，可以在这类任务上节约时间。

6

① https://devcenter.heroku.com/articles/custom-domains

第 7 章

整合前后端

提要：描述不同的部署方法、Chat 应用的最终版本和它的部署。

> 调试一段代码的难度是写出这段代码的两倍。因此，如果你的代码尽可能清楚，那么就不用费力的调试它。
>
> ——布莱恩 · W. 克尼汉[1]

现在，如果我们前端和后端的应用合并起来并且工作正常，那是极好的。下面是做这个事情的两种方式。

- 前端和后端不同域（Heroku 应用）：保证在使用 CORS 或者 JSONP 的时候没有跨域的问题。稍后会详细介绍这个方法。
- 相同域部署：确保 Node.js 处理前端静态文件和资源文件，这种方法在正式上线的应用中是不推荐的。

7.1 不同域部署

这是目前为止产品环境最好的实践。后端应用一般部署在 *http://app.*或者 *http://api.*子域名上。

为了使不同域的部署能正常工作，需要把受同一域限制的 AJAX 技术换成 JSONP：

```
var request = $.ajax({
  url: url,
  dataType: "jsonp",
  data: {...},
  jsonpCallback: "fetchData",
  type: "GET"
});
```

还有一种更好的方式，即在 Node.js 服务器应用返回前添加 `OPTIONS` 方法和特殊的响应头部信息，也叫 CORS：

---

① http://en.wikipedia.org/wiki/Brian_Kernighan

```
...
response.writeHead(200, {
  'Access-Control-Allow-Origin': origin,
  'Content-Type': 'text/plain',
  'Content-Length': body.length
});
...
```

或者：

```
...
res.writeHead(200, {
  'Access-Control-Allow-Origin': 'your-domain-name',
  ...
});
...
```

`OPTIONS` 方法所需的东西在 HTTP 访问控制（CORS）[1]里有定义。`OPTIONS` 请求可以通过下面的代码进行处理：

```
...
if (request.method == "OPTIONS") {
  response.writeHead("204", "No Content", {
    "Access-Control-Allow-Origin": origin,
    "Access-Control-Allow-Methods":
      "GET, POST, PUT, DELETE, OPTIONS",
    "Access-Control-Allow-Headers": "content-type, accept",
    "Access-Control-Max-Age": 10, // 秒
    "Content-Length": 0
  });
  response.end();
};
...
```

7.2 修改入口

此前，我们的前端应用使用 Parse.com 来作为后端应用的替代品。现在，我们切换回自己的后端。（无痛快速切换！）前端应用的源代码在 GitHub 上的 rpjs/board[2]目录里。

在 app.js 的开头，注释掉第一行本地运行的代码，或者把 URL 的值换成你的 Heroku 或 Windows Azure 后端应用的公共 URL：

```
// var URL = "http://localhost:5000/";
var URL = "http://your-app-name.herokuapp.com/";
```

如你所见，大部分 app.js 里的代码和文件夹结构保持完整，除了把 Parse.com 的模型和集合

① https://developer.mozilla.org/en-US/docs/HTTP_access_control

② https://github.com/azat-co/rpjs/tree/master/board

替换成了原来的 Backbone.js 的：

```
Message = Backbone.Model.extend({
  url: URL + "messages/create.json"
})
MessageBoard = Backbone.Collection.extend({
  model: Message,
  url: URL + "messages/list.json"
});
```

在这些地方 Backbone.js 查找 REST API 网址来与具体的集合和模型相一致。

下面是完整的 rpjs/board/app.js[①]文件源码：

```
/*
 这是一本关于 JavaScript 和 Node.js 的书，
 它将教你如何快速创建移动和 Web 应用，
 更多内容请访问：http://rapidprototypingwithjs.com
 */

// var URL = "http://localhost:5000/";
var URL = "http://your-app-name.herokuapp.com/";

require([
  'libs/text!header.html',
  'libs/text!home.html',
  'libs/text!footer.html'],

function (
headerTpl,
homeTpl,
footerTpl) {

  var ApplicationRouter = Backbone.Router.extend({
    routes: {
      "": "home",
      "*actions": "home"
    },
    initialize: function () {
      this.headerView = new HeaderView();
      this.headerView.render();
      this.footerView = new FooterView();
      this.footerView.render();
    },
    home: function () {
      this.homeView = new HomeView();
      this.homeView.render();
    }
  });

  HeaderView = Backbone.View.extend({
    el: "#header",
    templateFileName: "header.html",
```

① https://github.com/azat-co/rpjs/blob/master/board/app.js

```
  template: headerTpl,
  initialize: function () {};
  render: function () {
    $(this.el).html(_.template(this.template));
  }
});

FooterView = Backbone.View.extend({
  el: "#footer",
  template: footerTpl,
  render: function () {
    this.$el.html(_.template(this.template));
  }
});
Message = Backbone.Model.extend({
  url: URL + "messages/create.json"
})
MessageBoard = Backbone.Collection.extend({
  model: Message,
  url: URL + "messages/list.json"
});

HomeView = Backbone.View.extend({
  el: "#content",
  template: homeTpl,
  events: {
    "click #send": "saveMessage"
  },

  initialize: function () {
    this.collection = new MessageBoard();
    this.collection.bind("all", this.render, this);
    this.collection.fetch();
    this.collection.on("add", function (message) {
      message.save(null, {
        success: function (message) {
          console.log('saved ' + message);
        },
        error: function (message) {
          console.log('error');
        }
      });
      console.log('saved' + message);
    })
  },
  saveMessage: function () {
    var newMessageForm = $("#new-message");
    var username = newMessageForm.find('[name="username"]')
      .attr('value');
    var message = newMessageForm.find('[name="message"]')
      .attr('value');
    this.collection.add({
      "username": username,
      "message": message
    });
```

```
    },
    render: function () {
      console.log(this.collection)
      $(this.el).html(_.template(
        this.template,
        this.collection
      ));
    }
  });

  app = new ApplicationRouter();
  Backbone.history.start();
});
```

7.3　Chat 应用

Node.js 后端应用的源代码在 rpjs/node[1] GitHub 目录里，它的目录结构是这样的：

```
/node
        -web.js
        -Procfile
        -package.json
```

这里是 web.js 的代码，它实现了 CORS 的响应头：

```
/*
这是一本关于 JavaScript 和 Node.js 的书，
它将教你如何快速创建移动和 Web 应用，
更多内容请访问：http://rapidprototypingwithjs.com
*/

var http = require('http');
var util = require('util');
var querystring = require('querystring');
var mongo = require('mongodb');

var host = process.env.MONGOHQ_URL ||
  "mongodb://localhost:27017/board";
//MONGOHQ_URL=mongodb://user:pass@server.mongohq.com/db_name

mongo.Db.connect(host, function (error, client) {
  if (error) throw error;
  var collection = new mongo.Collection(client, 'messages');
  var app = http.createServer(function (request, response) {
    var origin = (request.headers.origin || "*");
    if (request.method == "OPTIONS") {
      response.writeHead("204", "No Content", {
        "Access-Control-Allow-Origin": origin,
        "Access-Control-Allow-Methods": "GET, POST, PUT, DELETE, OPTIONS",
        "Access-Control-Allow-Headers": "content-type, accept",
        "Access-Control-Max-Age": 10, // 秒
```

① https://github.com/azat-co/rpjs/tree/master/node

```
        "Content-Length": 0
      });
      response.end();
    }
    if (request.method === "GET" &&
      request.url === "/messages/list.json") {
      collection.find().toArray(function (error, results) {
        var body = JSON.stringify(results);
        response.writeHead(200, {
        'Access-Control-Allow-Origin': origin,
        'Content-Type': 'text/plain',
        'Content-Length': body.length
      });
      console.log("LIST OF OBJECTS: ");
      console.dir(results);
      response.end(body);
    });
  }
  if (request.method === "POST" &&
    request.url === "/messages/create.json") {
    request.on('data', function (data) {
      console.log("RECEIVED DATA:")
      console.log(data.toString('utf-8'));
      collection.insert(JSON.parse(data.toString('utf-8')),
      {safe: true}, function (error, obj) {
        if (error) throw error;
        console.log("OBJECT IS SAVED: ")
        console.log(JSON.stringify(obj))
        var body = JSON.stringify(obj);
        response.writeHead(200, {
              'Access-Control-Allow-Origin': origin,
          'Content-Type': 'text/plain',
          'Content-Length': body.length
        });
        response.end(body);
      })
    })

  };

  });
  var port = process.env.PORT || 5000;
  app.listen(port);
})
```

7.4 部署

为了方便起见，前端代码放在 rpjs/board[①]目录里，拥有 CORS 功能的后端应用在 rpjs/node[②]里。现在，你可能已经知道应该做什么了。作为参考，下面是把代码部署到 Heroku 上的步骤。

① https://github.com/azat-co/rpjs/tree/master/board

② https://github.com/azat-co/rpjs/tree/master/node

在 node 目录执行：

```
$ git init
$ git add .
$ git commit -am "first commit"
$ heroku create
$ heroku addons:add mongohq:sandbox
$ git push heroku master
```

复制并且粘贴地址到 board/app.js 文件，把它赋值给 URL 变量，然后在 board 目录里执行：

```
$ git init
$ git add .
$ git commit -am "first commit"
$ heroku create
$ git push heroku master
$ heroku open
```

7.5 同域部署

我们不推荐在生产环境中进行同域部署，因为静态资源文件使用像 nginx（非 Node.js I/O 引擎）这样的 Web 服务器更合适，并且分离 API 可以减少测试的复杂性，提高程序的健壮程度，更快的定位问题和监控。当然，同域部署可以用来测试、演示、开发环境或者微小型应用。

这是一个 Node.js 服务器静态应用的例子：

```
var http = require("http"),
  url = require("url"),
  path = require("path"),
  fs = require("fs"),
  port = process.argv[2] || 8888;

  http.createServer(function (request, response) {

  var uri = url.parse(request.url).pathname
    , filename = path.join(process.cwd(), uri);

  path.exists(filename, function (exists) {
    if(!exists) {
      response.writeHead(404, {
        "Content-Type": "text/plain"});
      response.write("404 Not Found\n");
      response.end();
      return;
    }

        if (fs.statSync(filename).isDirectory())
          filename += '/index.html';

    fs.readFile(filename, "binary",
```

```
      function (err, file) {
               if(err) {
                 response.writeHead(500,
                   {"Content-Type": "text/plain"});
                 response.write(err + "\n");
                 response.end();
                 return;
               }
         response.writeHead(200);
         response.write(file, "binary");
         response.end();
      });
   });
}).listen(parseInt(port, 10));

console.log("Static file server running at\n " +
  " => http://localhost:" + port + "/\nCTRL + C to shutdown");
```

注意

另外，更优雅的方式是使用 Node.js 框架，比如 Connect（http://www.senchalabs.org/connect/static.html）或者 Express（http://expressjs.com/guide.html），因为它们有专门用于 JS 和 CSS 资源的 static 中间件。

7

第 8 章

福利：Webapplog 上的文章

提要：一些关于 Node.js 里异步的本质的文章，通过 Mocha 使用 TDD；介绍 Express.js、Monk、Wintersmith、Derby 等框架和库。

> "不要害怕失败，你只需要做对一次。"
>
> ——安德鲁 · 豪斯顿[①]

为了方便本书的读者使用，在这章我们引入了一些 Webapplog.com 上关于 Node.js 的文章。Webapplog.com 是一个公开的关于 Web 开发的博客。

8.1 Node 里的异步

8.1.1 非阻塞I/O

与使用 Python 或者 Ruby 相比，使用 Node.js 最大的好处是有非阻塞 I/O 机制。为了说明这一点，我们以星巴克咖啡店里的队伍为例。假设每个人排队领取咖啡是一个任务，那么柜台后面所有的人和物，例如收银机、注册以及服务生，都像是服务器或者服务器应用。当我们要一杯普通的咖啡，比如 Pike Place、hot tea 或者 Earl Grey，服务生会制作它。整个队伍咖啡制作过程中都会等待，且每个人付适当的费用。

当然，上述那些饮品（众所周知是"耗时的瓶颈"）制作上都很简单，只需要倒出液体，就可以做好。但是 choco-mocha-frappe-latte-soy-decafs 怎么办？如果队列中每个人都决定购买这种耗时的饮料呢？这个队列会停滞不前，并且变得越来越长。咖啡店的经理必须增加更多的服务生，甚至自己开始收银。

这样不太好，对吧？但这个比喻生动地展现了除 Node.js 以外所有服务器端技术都会遭遇的情况，就像是真正的一家星巴克咖啡店。当你下单的时候，服务生把订单传递给别的雇员，你离开队列。队伍移动，处理器异步处理并且不会阻止队列。

---

① http://en.wikipedia.org/wiki/Drew_Houston

这就是 Node.js 能够在性能和扩展上打败同类产品（也许底层的 C++除外）的原因。使用 Node.js，你将不需要大量的 CPU 和服务器来处理负载。

8.1.2 异步编码方式

异步需要程序员有不同于自己所熟悉的 Python、PHP、C 或 Ruby 的思路。因为如果忘记在执行的时候返回正确的表达式，这非常容易引发错误。

下面这个简单的例子描述了这种情况：

```
var test = function (callback) {
  return callback();
  console.log('test') //不会被打印
}

var test2 = function (callback) {
  callback();
  console.log('test2') //第三个被打印
}

test(function () {
  console.log('callback1') //第一个被打印
  test2(function () {
  console.log('callback2') //第二个被打印
  })
});
```

如果我们不使用 `return callback()`，仅仅使用 `callback()`，字符串 `test2` 将会打印出来，`test` 不会。

```
callback1
callback2
test2
```

娱乐一下，我给 `callback2` 字符串添加了一个延迟 `setTimeout()`，现在顺序改变了：

```
var test = function (callback) {
  return callback();
  console.log('test') //不会被打印
}

var test2 = function (callback) {
  callback();
  console.log('test2') //第二个被打印
}

test(function () {
  console.log('callback1') //第一个被打印
  test2(function () {
    setTimeout(function () {
      console.log('callback2') //第三个被打印
    }, 100)
  })
});
```

打印：

```
callback1
tes2
callback2
```

最后的例子展示了两个并行运行的独立函数。较快的函数将会比较慢的那个更早结束。再回到星巴克的例子，这意味着，你可能会比队伍里排在你前面的人更快得到自己的饮料。对用户更友好，对编程也更好！

8.2　使用 Monk 迁移 MongoDB

最近，我们的一个高等用户抱怨他的 Storify[①]账户不能登录。我们检查了产品数据库，发现可能有人用这个账户的用户名和密码登录并且恶意删除了它。多亏了伟大的 MongoHQ 服务，我们得以在 15 分钟内恢复数据库。进行这个操作有两种选择：

(1) Mongo shell 脚本；
(2) Node.js 程序。

因为这个 Storify 用户账户删除了所有相关的对象，如验证、关系（关注、被关注）、喜欢、故事等，我们决定使用第二种方案。它工作得非常棒，这里有一个简洁版，你可以在 MongoDB 迁移中使用（也放在了 gist.github.com/4516139[②]里）。

现在我们来加载所有的模块：Monk[③]、Progress[④]、Async[⑤]以及 MongoDB：

```
var async = require('async');
var ProgressBar = require('progress');
var monk = require('monk');
var ObjectId = require('mongodb').ObjectID;
```

顺便说一下，Monk 是由 LeanBoost[⑥]开发，它是 Node.js 里使用 MongoDB 的一个简易且对用户友好的封装。

Monk 使用下面的连接字符串格式：

```
username:password@dbhost:port/database
```

① http://storify.com
② https://gist.github.com/4516139
③ https://github.com/LearnBoost/monk
④ https://github.com/visionmedia/node-progress
⑤ https://github.com/caolan/async
⑥ https://www.learnboost.com/

创建下面的对象：

```
var dest = monk('localhost:27017/storify_localhost');
var backup = monk('localhost:27017/storify_backup');
```

我们需要知道要保存的对象的 ID：

```
var userId = ObjectId(YOUR-OBJECT-ID);
```

这是人工输入的 `restore()` 函数，它可以通过指定特定的查询重复使用已经保存的对象（更多关于 MongoDB 查询的内容，请查看文章"Querying 20M-Record MongoDB Collection"[1]）。如果想调用它，只需要把集合的名字以一个字符串传入，比如`"stories"`，并且从主对象里取值关联对象，比如`{userId:user.id}`。进度条可以在终端里形象展示当前进度：

```
var restore = function (collection, query, callback) {
  console.info('restoring from ' + collection);
  var q = query;
  backup.get(collection).count(q, function (e, n) {
    console.log('found ' + n + ' ' + collection);
    if (e) console.error(e);
    var bar = new ProgressBar('[:bar] :current/:total'
      + ':percent :etas'
      , { total: n - 1, width: 40 })
    var tick = function (e) {
      if (e) {
        console.error(e);
        bar.tick();
      }
      else {
        bar.tick();
      }
      if (bar.complete) {
        cinsole.log();
        console.log('restoring ' + collection + ' is completed');
        callback();
      }
    };
    if (n > 0) {
      console.log('adding ' + n + ' ' + collection);
      backup.get(collection).find(q, {
        stream: true
      }).each(function (element) {
        dest.get(collection).insert(element, tick);
      });
    } else {
      callback();
    }
  });
}
```

现在我们使用 `async` 来调用上面提到的 `restore()` 函数：

① http://www.webapplog.com/querying-20m-record-mongodb-collection/

```
async.series({
  restoreUser: function (callback) { // 导入用户元素
    backup.get('users').find({_id: userId}, {
      stream: true, limit: 1
    }).each(function (user) {
      dest.get('users').insert(user, function (e) {
        if (e) {
          console.log(e);
        }
        else {
          console.log('resored user: ' + user.username);
        }
        callback();
      });
    });
  },

  restoreIdentity: function (callback) {
    restore('identities', {
      userid: userId
    }, callback);
  },
  restoreStories: function (callback) {
    restore('stories', {authorid: userId}, callback);
  }

  }, function (e) {
  console.log();
  console.log('restoring is completed!');
  process.exit(1);
});
```

完整的代码在gist.github.com/4516139[①]里，下面也是：

```
var async = require('async');
var ProgressBar = require('progress');
var monk = require('monk');
var ms = require('ms');
var ObjectId = require('mongodb').ObjectID;

var dest = monk('localhost:27017/storify_localhost');
var backup = monk('localhost:27017/storify_backup');

var userId = ObjectId(YOUR - OBJECT - ID);
// monk 会自动分配，但是我们需要用它来进行查询

var restore = function (collection, query, callback) {
  console.info('restoring from ' + collection);
  var q = query;
  backup.get(collection).count(q, function (e, n) {
    console.log('found ' + n + ' ' + collection);
    if (e) console.error(e);
```

① https://gist.github.com/4516139

```
    var bar = new ProgressBar(
      '[:bar] :current/:total :percent :etas',
      { total: n - 1, width: 40 })
    var tick = function (e) {
      if (e) {
        console.error(e);
        bar.tick();
      }
      else {
        bar.tick();
      }
      if (bar.complete) {
        console,log();
        console.log('restoring ' + collection + ' is completed');
        callback();
      }
    };
    if (n > 0) {
      console.log('adding ' + n + ' ' + collection);
      backup.get(collection).find(q, { stream: true })
        .each(function (element) {
        dest.get(collection).insert(element, tick);
      });
    } else {
      callback();
    }
  });
}

async.series({
  restoreUser: function (callback) {// 导入用户元素
    backup.get('users').find({_id: userId}, {
      stream: true,
      limit: 1 })
      .each(function (user) {
      dest.get('users').insert(user, function (e) {
        if (e) {
          console.log(e);
        }
        else {
          console.log('resored user: ' + user.username);
        }
        callback();
      });
    });
  },

  restoreIdentity: function (callback) {
    restore('identities', {
      userid: userId
    }, callback);
  },

  restoreStories: function (callback) {
```

```
    restore('stories', {authorid: userId}, callback);
  }

  }, function (e) {
  console.log();
  console.log('restoring is completed!');
  process.exit(1);
});
```

运行 `npm insall/npm update` 并且修改硬编码的数据库值来运行它。

8.3 在 Node.js 里使用 Mocha 实践 TDD

8.3.1 谁需要使用测试驱动的开发

设想一下你要在一个已经存在的接口上实现一个复杂的功能，比如在评论上添加一个“like”（赞）按钮。在没有测试的情况下，必须人工创建用户，登录，创建文章，创建另一个用户，登录，赞这个文章。很令人厌烦吧？如果你需要重复这个步骤10到20次来发现和修复某些bug呢？如果你新添加的功能破坏了已经有的功能，而且在没有测试的情况下，6个月后才发现这一漏洞，怎么办呢？

不要为一次性的代码写测试，但是针对主代码，请养成测试驱动的习惯。只需要在开始的时候花点儿时间，你和你的团队稍后就可以节约很多时间并且在发布的时候更有自信。测试驱动开发真的是益处多多的好事情！

8.3.2 快速开始指南

请按照这个快速指南使用 Mocha①来设置测试驱动开发环境。

执行下面的命令，全局安装 Mocha②：

```
$ sudo npm install -g mocha
```

我们还要使用另外两个库：LearnBoost③的 Superagent④和 expect.js⑤。安装它们，在项目目录里使用 NPM 命令⑥：

```
$ npm install superagent
$ npm install expect.js
```

---

① http://visionmedia.github.com/mocha/
② http://visionmedia.github.com/mocha/
③ https://github.com/LearnBoost
④ https://github.com/visionmedia/superagent
⑤ https://github.com/LearnBoost/expect.js/
⑥ https://npmjs.org/

打开一个新的.js 文件，输入：

```
var request = require('superagent');
var expect = require('expect.js');
```

目前我们已经载入了两个库。测试套件的结构看起来是这样的：

```
describe('Suite one', function () {
  it(function (done) {
  ...
  });
  it(function (done) {
  ...
  });
});
describe('Suite two', function () {
  it(function (done) {
  ...
  });
});
```

在这个封闭包里，我们写一个针对我们的服务器的请求，假设它在 localhost:8080[①]：

```
...
it(function (done) {
  request.post('localhost:8080').end(function (res) {
    //TODO 检查响应是否正确
  });
});
...
```

Expect 提供了用来检查返回是否正确的便捷函数：

```
...
expect(res).to.exist;
expect(res.status).to.equal(200);
expect(res.body).to.contain('world');
...
```

最后，我们需要添加 done()调用，告诉 Mocha，这个异步测试已经完成。我们第一个测试的完整代码如下：

```
var request = require('superagent');
var expect = require('expect.js');

describe('Suite one', function () {
 it(function (done) {
   request.post('localhost:8080').end(function (res) {
    expect(res).to.exist;
    expect(res.status).to.equal(200);
    expect(res.body).to.contain('world');
    done();
  });
 });
```

① http://localhost:8080

如果我们想在请求前处理，可以添加 before 和 beforeEach 钩子，顾名思义，它们在每一个测试（或测试套件）运行前执行：

```
before(function () {
  //TODO 初始化数据库
});
describe('suite one ', function () {
  beforeEach(function () {
    //TODO 登录测试用户
  });
  it('test one', function (done) {
  ...
  });
});
```

请注意，before 和 beforeEach 可以放在 describe 里，也可以放在外面。运行这个测试，简单执行：

```
$ mocha test.sj
```

使用不同的报告类型：

```
$ mocha test.js -R list
$ mocah test.js -R spec
```

8.4　Wintersmith：静态网站生成器

针对这本书的单页网站 rapidprototypingwithjs.com①，我使用 Wintersmith②来学习并快速启动了一些东西。Wintersmith 是一个 Node.js 静态网站生成器。它的可扩展和方便的部署带给了我极大的震憾。另外，这里还有几个我最喜欢的工具，比如 Markdown③、Jade 和 Underscrore④。

为什么选择静态网站生成器

这里有一个文章解释了为什么一般情况下使用静态网站生成器是好主意：An Introduction to Static Site Generators⑤。它依据的是下面几样重要的东西。

模板

可以使用诸如 Jade⑥的模板引擎。Jade 使用空格来组织级联元素，它的语法和 Ruby on Rail's 的 Haml 标记很像。

① http://rapidprototypingwithjs.com
② http://jnordberg.github.com/wintersmith/
③ http://daringfireball.net/projects/markdown/
④ http://underscorejs.org/
⑤ http://www.mickgardner.com/2012/12/an-introduction-to-static-site.html
⑥ https://github.com/visionmedia/jade

Markdown

我曾经从自己某本书的介绍章节复制 markdown 文本，并且没做任何修改直接使用它。Wintersmith 使用了 marked[1]作为 Markdown 解析器。更多关于为什么 Markdown 是很棒的，请参考我的文章：Markdown Goodness[2]。

简单的部署

所需的东西是 HTML、CSS 和 JavaScript，所以你只需要使用一个 FTP 客户端上传它们，比如 Panic 的 Transmit[3]或 Cyberduck[4]。

基本服务

由于任何静态 Web 服务器都可以正常工作，没必要使用 Heroku 或者 Nodejitsu 私有云，甚至 PHP/MySQL 托管服务。

性能

没有数据库调用，没有服务器端 API 调用，也不会有 CPU 或内存过载。

灵活性

Wintersmith 可以为内容和模板加载插件，你也可以写一个自己的插件[5]。

8.4.1　开始使用Wintersmith

github.com/jnordberg/wintersmith[6]这里有快速指南。

全局安装 Wintersmith，使用`-g`参数和`sudo`来运行 NPM：

```
$ sudo npm install wintersmith -g
```

使用默认的博客模板来运行：

```
$ wintersmith new <path>
```

或者使用一个空网站：

```
$ wintersmith new <path> -template basic
```

或者使用快捷方式：

```
$ wintersmith new <path> -T basic
```

---

① https://github.com/chjj/marked

② http://www.webapplog.com/markdown-goodness/

③ http://www.panic.com/transmit/

④ http://cyberduck.ch/

⑤ https://github.com/jnordberg/wintersmith#content-plugins

⑥ https://github.com/jnordberg/wintersmith

类似 Ruby on Rails 支架，Wintersmith 将生成一个带有内容和模板目录的基本框架。预览这个网站，运行下面的命令：

```
$ cd <path>
$ wintersmith preview
$ open http://localhost:8080
```

大多数的改变在预览模式下可以自动更新，config.json 文件[1]除外。图像、CSS、JavaScript 和其他文件会移动到 contents 文件夹。Wintersmith 生成器用的是下面的逻辑：

(1) 查找 contents 目录里的*.md 文件；

(2) 阅读 metadata[2]，例如模板名；

(3) 处理每一个*.md 文件里后缀名为*.jade 的模板[3]里的 metadata。

当你创建好了自己的静态网站，运行：

```
$ wintersmith build
```

8.4.2 其他静态网站生成器

这里还有一些别的 Node.js 静态网站生成器：

- Docpad[4]
- Blacksmith[5]
- Scotch[6]
- Wheat[7]
- Petrify[8]

关于这些静态网站生成器更详细概述参见：Node.js Based Static Site Generators[9]。

其他语言如 Rails 和 PHP 的生成器，请查看："按 Github 关注数排序的静态网站生成器列表"[10]和"mother of all site generator lists"[11]。

---

① https://github.com/jnordberg/wintersmith#config
② https://github.com/jnordberg/wintersmith#the-page-plugin
③ https://github.com/jnordberg/wintersmith#templates
④ https://github.com/bevry/docpad#readme
⑤ https://github.com/flatiron/blacksmith
⑥ https://github.com/techwraith/scotch
⑦ https://github.com/creationix/wheat
⑧ https://github.com/caolan/petrify
⑨ http://blog.bmannconsulting.com/node-static-site-generators/
⑩ https://gist.github.com/2254924
⑪ http://nanoc.stoneship.org/docs/1-introduction/#similar-projects

8.5 Express.js 教程：使用 Monk 和 MongoDB 的简单 REST API 应用

使用Express.js和Monk构建REST API应用

这个应用是 mongoui[1]的开始，它是使用 Node.js 为 MongoDB 编写的相当于 phpMyAdmin 的应用，目的是提供一个友好的管理员界面。它有点像 Parse.com、Firebase.com、MongoHQ[2]）和 MongoLab[3]，但是不会把它变成特定的服务。为什么每次查找用户信息我们需要输入 `db.users.findOne({'_id':ObjectId('...')})`？可供选用的 MongoHub[4]Mac 应用也拥有同样的功能，而且是免费的，但使用起来比较笨重，并且不是基于 Web 的。

Ruby 的热衷者乐于把 Express 和 Sinatra[5]框架进行对比。在创建应用的方式上它们都很灵活。应用路由设置代码类似，即 `app.get('/products/:id',showProduct);`。Express.js 当前的版本是 3.1，为了辅助 Express，我们使用 Monk[6]模块。

我们将使用 NPM（Node Package Manager）[7]，它在 Node.js 安装时已经附带安装。如果你还没有安装它，可以在 npmjs.org[8]下载。

创建一个新文件夹和 NPM 配置文件 package.json，在它里面写入下面的内容：

```
{
  "name": "mongoui",
  "version": "0.0.1",
  "engines": {
    "node": ">= v0.6"
  },
  "dependencies": {
    "mongodb": "1.2.14",
    "monk": "0.7.1",
    "express": "3.1.0"
  }
}
```

现在运行 `npm install` 下载并安装模块到 node_module 目录。如果一切正常，在 node_module 目录里可以看到很多目录。为了保持简洁，我们把应用中所有代码都放在一个 index.js 文件里：

---

① http://gitbhub.com/azat-co/mongoui
② http://mongohq.com
③ http://mongolab.com
④ http://mongohub.todayclose.com/
⑤ http://www.sinatrarb.com/
⑥ https://github.com/LearnBoost/monk
⑦ http://npmjs.org
⑧ http://npmjs.org

```
var mongo = require('mongodb');
var express = require('express');
var monk = require('monk');
var db = monk('localhost:27017/test');
var app = new express();

app.use(express.static(__dirname + '/public'));
app.get('/', function (req, res) {
  db.driver.admin.listDatabases(function (e, dbs) {
    res.json(dbs);
  });
});
app.get('/collections', function (req, res) {
  db.driver.collectionNames(function (e, names) {
    res.json(names);
  })
});
app.get('/collections/:name', function (req, res) {
  var collection = db.get(req.params.name);
  collection.find({}, {limit: 20}, function (e, docs) {
    res.json(docs);
  })
});
app.listen(3000)
```

让我们把代码拆开来一点一点分析，首先是引入模块声明：

```
var mongo = require('mongodb');
var express = require('express');
var monk = require('monk');
```

数据库和 Express 应用的实例化：

```
var db = monk('localhost:27017/test');
var app = new express();
```

告诉 Express 应用从 public 目录加载和服务器静态文件：

```
app.use(express.static(__dirname + '/public'));
```

首页，也叫根路由的设置：

```
app.get('/', function (req, res) {
  db.driver.admin.listDatabases(function (e, dbs) {
    res.json(dbs);
  });
});
```

get()函数需要两个参数：字符串和函数。字符串可以包含斜线和冒号，比如 product/:id。函数必须有两个参数：请求和响应。请求包含所有的诸如查询字符串、会话和首部的信息，响应是需要返回的结果。在这个例子里，我们调用 res.json()函数来返回。

如你所料，db.driver.admin.listDatabases()以异步的方式返回数据库列表。

其他两个路由以和 get() 相似的形式设置：

```
app.get('/collections', function (req, res) {
  db.driver.collectionNames(function (e, names) {
    res.json(names);
  })
});
app.get('/collections/:name', function (req, res) {
  var collection = db.get(req.params.name);
  collection.find({}, {limit: 20}, function (e, docs) {
    res.json(docs);
  })
});
```

Express 非常便于支持其他 HTTP 动作，比如 post 和 update。在设置 post 路由的时候，我们这样写：

```
app.post('product/:id',function(req,res) {...});
```

Express 也支持中间件。中间件是一个请求处理函数，它的三个参数为：request、response 和 next。比如：

```
app.post('product/:id',
  authenticateUser,
  validateProduct,
  addProduct
);

function authenticateUser(req, res, next) {
  //通过检查 req.session 来验证用户
  next();
}

function validateProduct(req, res, next) {
   //校验提交的数据
   next();
}

function addProduct(req, res) {
  //保存数据到数据库
}
```

validateProduct 和 authenticateProduct 是中间件。在大项目中人们通常把它们放到单独的文件里。

另一个在 Express 应用里设置中间件的方式是使用 use() 函数。例如，之前我们为静态资源所做的：

```
app.use(express.static(__dirname + '/public'));
```

我们同样可以这样处理错误：

```
app.use(errorHandler);
```

假定，我们已经安装了 MongoDB，这个应用会连接到它（localhost:27017[①]）并且展示集合的名字和它里面的项目。打开 mongo 服务器：

```
$ mongod
```

运行这个应用（保持 mongod 终端呈打开状态）：

```
$ node
```

或者：

```
$ node index.js
```

查看应用的工作状态，使用 Chrome 打开 localhost:3000[②]，使用 JSONViewer[③]进行扩展（它可以以更友好的方式展示 JSON）。

8.6 Express.js 教程：参数、错误处理及其他中间件

8.6.1 请求处理函数

Express.js 是一个 node.js 框架，相比于其他，它提供了组织路由的方式。每一个路由通过把 URL（也可以使用正则表达式）作为第一个参数调用应用对象上的函数。例如：

```
app.get('api/v1/stories/', function (res, req) {
  ...
})
```

或者用 POST 方法：

```
app.post('api/v1/stories/' function (req, res) {
  ...
})
```

不用多说，DELETE 和 PUT 方法也被给予很好的支持[④]。

我们传递给 get() 或者 post() 方法的回调叫做请求处理函数，因为它们会接收并处理请求（req），然后写入到响应（res）对象。例如：

```
app.get('/about', function (req, res) {
  res.send('About Us: ...');
});
```

---

① http://localhost:27017

② http://localhost:3000

③ https://chrome.google.com/webstore/detail/jsonview/chklaanhfefbnpoihckbnefhakgolnmc?hl=en

④ http://expressjs.com/api.html#app.VERB

我们可以使用多个请求处理函数，因此它们的名字叫中间件。它们接收第三个参数，即 next，调用它（next()）就会顺序执行下一个请求处理函数：

```
app.get('/api/v1/stories/:id', function (req, res, next) {
  //进行授权验证
  //如果没有通过验证或者有错误，返回 next(error);
  //如果通过并且没有错误
  return next();
}), function (req, res, next) {
  //获取 id 并且从数据库里取回数据
  //如果没有错误，保存 story 到请求对象上
  req.story = story;
  return next();
}), function (req, res) {
  //输出数据库查询的结果
  res.send(res.story);
});
```

我们需要用 URL 字符串里的 ID 参数查询它在数据库里对应的匹配项。

8.6.2 参数处理中间件

参数是请求 URL 里的对应值。如果我们没有 Express.js 或者类似的库而使用核心 Node.js 模块，我们必须使用 HTTP.request[①]对象，然后使用 require('querystring').parse(url)或 require('url').parse(url, true)这样的处理。

多亏了 Connect[②]框架和 VisionMedia[③]方面的人员，Express.js 得以以中间件形式来支持参数，处理错误，并且具备其他很多重要功能。下面是我们如何在应用里处理参数：

```
app.param('id', function (req, res, next, id) {
  //使用 id 做一些事情
  //保存 id 或者其他信息到请求对象中
  //完成后调用 next
  next();
});

app.get('/api/v1/stories/:id', function (req, res) {
  //参数处理中间件将在此之前执行
  //我们期待请求对象上已经有了需要的信息
  //输出一些东西
  res.send(data);
});
```

例如：

```
app.param('id', function (req, res, next, id) {
  req.db.get('stories').findOne({_id: id}, function (e, story) {
```

① http://nodejs.org/api/http.html#http_http_request_options_call back

② http://www.senchalabs.org/connect/

③ https://github.com/visionmedia/express

```
    if (e) return next(e);
    if (!story) return next(new Error('Nothing is found'));
    req.story = story;
    next();
  });
});

app.get('/api/v1/stories/:id', function (req, res) {
  res.send(req.story);
});
```

或者我们使用多个请求处理函数，原理也一样：可以预期获取 req.story 对象或者一个之前执行里抛出的错误，因此可以抽象出获取参数及其各自对象的通用代码/逻辑：

```
app.get('/api/v1/stories/:id', function (req, res, next) {
  //授权验证
  }),
  //已经有对象在 req.story，所以这里不需要做什么
  function (req, res) {
  //输出数据库查询的结果
  res.send(story);
});
```

登录校验和输入过滤也非常适合放在中间件里。

param()函数非常酷，因为它可以合并不同的参数，比如：

```
app.get('/api/v1/stories/:storyId/elements/:elementId',
  function (req, res) {
    res.send(req.element);
  }
);
```

8.6.3　错误处理

错误处理通常贯穿于整个程序生命周期，所以它最好也是以一个中间件的形式存在。它有相同的参数，只是多了一个 error：

```
app.use(function (err, req, res, next) {
  //日志记录和用户友好的错误消息输出
  res.send(500);
})
```

事实上，响应可以是任何东西，如下。

JSON 字符串

```
app.use(function (err, req, res, next) {
  //日志记录和用户友好的错误消息输出
  res.send(500, {status: 500,
    message: 'internal error',
    type: 'internal'}
```

```
  );
})
```

纯文本信息

```
app.use(function (err, req, res, next) {
  //日志记录和用户友好的错误消息输出
  res.send(500, 'internal server error');
})
```

错误页

```
app.use(function (err, req, res, next) {
  //日志记录和用户友好的错误消息输出
  //假设模板引擎已经添加过
  res.render('500');
})
```

跳转到一个错误页

```
app.use(function (err, req, res, next) {
  //日志记录和用户友好的错误消息输出
  res.redirect('/public/500.html');
})
```

错误 HTTP 响应码（401、400、500 等）

```
app.use(function (err, req, res, next) {
  //日志记录和用户友好的错误消息输出
  res.end(500);
})
```

顺便提一下，日志也可以抽象成一个中间件。

在请求处理函数或者中间件里触发错误，只需要这样的调用：

```
next(error)
```

或者：

```
next(new Error('Something went wrong :-(');
```

你也可以使用多个错误处理函数，并使用具名函数，而不是使用匿名函数，Express.js Error handling guide[①]里有相关的例子。

8.6.4 其他中间件

除了提取参数，它还可用于其他很多情况，比如登录校验、错误处理、会话以及输出等。

res.json()就是它们中的一个。它把 JavaScript/Node.js 对象友好地转化为 JSON，例如：

① http://expressjs.com/guide.html#error_handling

```
app.get('/api/v1/stories/:id', function (req, res) {
  res.json(req.story);
});
```

在 req.story 是数组或者对象的时候等同于：

```
app.get('/api/v1/stories/:id', function (req, res) {
  res.send(req.story);
});
```

或者：

```
app.get('api/v1/stories/:id', function (req, res) {
  res.set({
    'Content-Type': 'application/json'
  });
  res.send(req.story);
});
```

8.6.5　抽象

中间件很灵活，可以使用匿名或者具名函数，最好的方式是把请求处理函数按功能抽象到单独的外部模块里：

```
var stories = require.('./routes/stories');
var elements = require.('./routes/elements');
var users = require.('./routes/users');
...
app.get('/stories/,stories.find);
app.get('/stories/:storyId/elements/:elementId', elements.find);
app.put('/users/:userId', users.update);
```

routes/stories.js：

```
module.exports.find = function(req,res, next) {
};
```

routes/elements.js：

```
module.exports.find = function(req,res,next){
};
```

routes/users.js：

```
module.exports.update = function(req,res,next){
};
```

我们可以使用一些函数式编程的技巧，比如：

```
function requiredParamHandler(param) {
  //使用 param 做一些事情，比如
  //检查它是否出现在请求字符串中
```

```
    return function (req, res, next) {
      //使用 param，比如如果 token 有效的，调用 next()
      next();
    });
  }

  app.get('/api/v1/stories/:id',
    requiredParamHandler('token'),
    story.show
  );

  var story = {
    show: function (req, res, next) {
      //一些逻辑处理，比如限制需要输出的字段
      return res.send();
    }
  }
```

如你所见，中间件在组织代码时是非常有用的原则。最佳实践是保持路由程序简洁精短，把所有的逻辑代码移到相对应的外部模块或者文件。这么做之后，当你需要时，重要的服务器配置参数会整齐地出现在一个地方。

8.7 使用 Node.js 和 MongoDB 通过 Mongoskin 和 Express.js 构建 JSON REST API 服务器

这个教程会带你使用 Mocha[①]和 Super Agent[②]库写测试，然后使用测试驱动的方式用 Express.js[③]、MongoDB[④]的 Mongoskin[⑤]库开发一个免费的 Node.js[⑥] JSON REST API 服务器。在这个 REST API 服务器上，我们将进行创建、更新、移除以及删除操作（CRUD），通过使用 `app.param()`和 `app.use()`方法实践 Express.js 的中间件[⑦]。

8.7.1 测试覆盖率

在开始做事情之前，我们写点功能测试，便于之后向我们的 REST API 服务器发送 HTTP 测试。如果你已经知道怎么使用 Mocha[⑧]或者想直接跳跃阅读了解 Express.js 应用是怎么实现的，完全没问题。你也可以使用终端的 CURL 命令来测试。

---

① http://visionmedia.github.io/mocha/

② http://visionmedia.github.io/superagent/

③ http://expressjs.com/

④ http://www.mongodb.org/

⑤ https://github.com/kissjs/node-mongoskin

⑥ http://nodejs.org

⑦ http://expressjs.com/api.html#middleware

⑧ http://visionmedia.github.io/mocha/

假设我们已经安装了Node.js、NPM[1]和MongoDB，那么创建一个新的目录（或者你已经写过测试了，就使用之前的目录）：

```
mkdir rest-api
cd rest-api
```

我们将要使用Mocha[2]、Exppect.js[3]和Super Agent[4]库。为了安装它们，请在项目目录里运行如下命令：

```
$ npm install mocha
$ npm install expect.js
$ npm install superagent
```

现在在相同的目录里创建一个express.test.js文件，它将会有6个测试套件：

- 创建一个新对象；
- 通过ID获取一个对象；
- 获取整个集合；
- 通过ID更新一个对象；
- 通过ID检查一个更新的对象；
- 通过ID删除某个对象。

使用Super Agent的链式调用函数，HTTP请求简直就是小菜一碟儿，我们将会把它放在每一个测试套件里。下面是完整的express.test.js的源代码：

```
var superagent = require('superagent')
var expect = require('expect.js')

describe('express rest api server', function () {
  var id

  it('post object', function (done) {
    superagent.post('http://localhost:3000/collections/test')
      .send({ name: 'John'
        , email: 'john@rpjs.co'
      })
      .end(function (e, res) {
        // console.log(res.body)
        expect(e).to.eql(null)
        expect(res.body.length).to.eql(1)
        expect(res.body[0]._id.length).to.eql(24)
        id = res.body[0]._id
        done()
      })
  })
```

① http://npmjs.org

② http://visionmedia.github.io/mocha/

③ https://github.com/LearnBoost/expect.js/

④ http://visionmedia.github.io/superagent/

```
it('retrieves an object', function (done) {
  superagent.get('http://localhost:3000/collections/test/' + id)
    .end(function (e, res) {
      // console.log(res.body)
      expect(e).to.eql(null)
      expect(typeof res.body).to.eql('object')
      expect(res.body._id.length).to.eql(24)
      expect(res.body._id).to.eql(id)
      done()
    })
})

it('retrieves a collection', function (done) {
  superagent.get('http://localhost:3000/collections/test')
    .end(function (e, res) {
      // console.log(res.body)
      expect(e).to.eql(null)
      expect(res.body.length).to.be.above(1)
      expect(res.body.map(function (item) {
        return item._id
      })).to.contain(id)
      done()
    })
})

it('updates an object', function (done) {
  superagent.put('http://localhost:3000/collections/test/' + id)
    .send({name: 'Peter'
      , email: 'peter@yahoo.com'})
    .end(function (e, res) {
      // console.log(res.body)
      expect(e).to.eql(null)
      expect(typeof res.body).to.eql('object')
      expect(res.body.msg).to.eql('success')
      done()
    })
})

it('checks an updated object', function (done) {
  superagent.get('http://localhost:3000/collections/test/' + id)
    .end(function (e, res) {
      // console.log(res.body)
      expect(e).to.eql(null)
      expect(typeof res.body).to.eql('object')
      expect(res.body._id.length).to.eql(24)
      expect(res.body._id).to.eql(id)
      expect(res.body.name).to.eql('Peter')
      done()
    })
})

it('removes an object', function (done) {
  superagent.del('http://localhost:3000/collections/test/' + id)
    .end(function (e, res) {
      // console.log(res.body)
```

```
            expect(e).to.eql(null)
            expect(typeof res.body).to.eql('object')
            expect(res.body.msg).to.eql('success')
            done()
        })
    })
})
```

运行测试，我们可以使用`$ mocha express.test.js`命令。

8.7.2　依赖

在这个教程里，我们将要使用Mongoskin[①]，它是一个MongoDB库，比原生的MongoDB驱动[②]更好用。另外，Mongoskin比Mongoose的量级更轻，Mongoose的schema更少。更多信息可以查看Mongoskin comparison blurb[③]。

Express.js[④]是Node.js核心模块HTTP模块[⑤]的包装器。它在Connect[⑥]中间件的基础上构建而来，提供了大量的方便。有些人会把它和Ruby的Sinatra框架对比，它们同样开放和方便配置。

如果在上一节测试覆盖率中你已经创建了`rest-api`目录，可以简单的运行如下命令来为应用安装模块：

```
npm install express
npm install mongoskin
```

8.7.3　实现

第一件事，定义我们的依赖：

```
var express = require('express')
  , mongoskin = require('mongoskin')
```

在版本3.*x*之后，Express提高了实例化的效率，也就是说下面的一行代码会返回一个服务器对象：

```
var app = express()
```

为了取出请求体里的参数，我们将使用`bodyParser()`中间件，它看起来更像是个配置语句：

```
app.use(express.bodyParser())
```

---

① https://github.com/kissjs/node-mongoskin

② https://github.com/mongodb/node-mongodb-native

③ https://github.com/kissjs/node-mongoskin#comparation

④ http://expressjs.com/

⑤ http://nodejs.org/api/http.html

⑥ https://github.com/senchalabs/connect

中间件（可以是这种形式[1]，也可以是其他形式[2]）是 Express.js 和 Connect[3]里一个强大且方便的代码组织和重用模式。

由于使用了 bodyParser()方法，所以我们从 HTTP 请求里分析出 body 对象的痛苦减轻了，Mongoskin 使得只需要使用一行代码就可以连接到 MongoDB 数据库：

```
var db = mongoskin.db('localhost:27017/test', {safe:true});
```

注意

如果你想连接到一个远程数据库，比如 MongoHQ[4]实例，需要提供你的用户名、密码、主机和端口值。下面是一种这样的 URI 字符串格式：*mongodb://[username:password@]host1[:port1]*。

app.param()方法是另外一个 Express.js 中间件。它基本上就是：每一次请求处理里有一个值的时候做一些事情。当请求字符中有 collectonName 的时候，我们选择了一个特殊的集合。稍后你会在路由里看到它：

```
app.param('collectionName',
  function (req, res, next, collectionName) {
    req.collection = db.collection(collectionName)
    return next()
  }
)
```

为了用户友好体验，我们给根路由添加一个反馈消息：

```
app.get('/', function (req, res) {
  res.send('please select a collection, e.g., /collections/messages')
})
```

现在我们真正的工作开始了。下面是我们如何以 _id 排序，限制 10 个获取项目的列表：

```
app.get('/collections/:collectionName',
  function (req, res) {
    req.collection
      .find({},
        {limit: 10, sort: [['_id', -1]]}
      ).toArray(function (e, results) {
        if (e) return next(e)
        res.send(results)
      }
    )
  }
)
```

---

① http://expressjs.com/api.html#app.use

② http://expressjs.com/api.html#middleware

③ https://github.com/senchalabs/connect

④ https://www.mongohq.com/home

你有没有注意到URL字符串里的:collectionName？这个和之前我们使用的app.param()中间件会给我们一个指向数据库里特定集合的req.collection。

这个对象创建略微简单点，因为我们只需要把所有的东西插入 MongoDB（这种方法也叫JSON REST API）：

```
app.post('/collections/:collectionName', function (req, res) {
  req.collection.insert(req.body, {}, function (e, results) {
    if (e) return next(e)
    res.send(results)
  })
})
```

单个对象获取函数比find()更快，但是它们的接口不同（它们直接返回对象而不是指针），所以要小心使用。另外，我们通过Express.js的魔力req.params.id从路径的:id获取ID：

```
app.get('/collections/:collectionName/:id', function (req, res) {
  req.collection.findOne({_id: req.collection.id(req.params.id)},
    function (e, result) {
      if (e) return next(e)
      res.send(result)
    }
  )
})
```

PUT请求的处理更加有趣，因为update()不会返回传入的对象，而是返回它修改的对象的数量。

另外{$set:req.body}也是一个特殊的MongoDB操作（操作符是以美元符开着），它设置值。

第二个{safe:true, multi:false}参数是一个带有选项的对象，它告诉MongoDB在运行回调函数之前等待执行并且只处理第一个项目：

```
app.put('/collections/:collectionName/:id', function (req, res) {
  req.collection.update({_id: req.collection.id(req.params.id)},
    {$set: req.body},
    {safe: true, multi: false},
    function (e, result) {
      if (e) return next(e)
      res.send((result === 1) ? {msg: 'success'} : {msg: 'error'})
    }
  )
})
```

最后，DELETE方法也输出一个自定义的JSON消息：

```
app.del('/collections/:collectionName/:id', function (req, res) {
  req.collection.remove({_id: req.collection.id(req.params.id)},
    function (e, result) {
      if (e) return next(e)
      res.send((result === 1) ? {msg: 'success'} : {msg: 'error'})
    }
  )
})
```

注意

delete 是 JavaScript 里的一个操作符，Express.js 使用 app.del 替代它。

最后一行，在端口 3000 开启服务器，开始监听：

```
app.listen(3000)
```

如果有些时候它不正常工作，请参考完整的 express.js 代码：

```
var express = require('express')
  , mongoskin = require('mongoskin')

var app = express()
app.use(express.bodyParser())

var db = mongoskin.db('localhost:27017/test', {safe: true});

app.param('collectionName',
  function (req, res, next, collectionName) {
    req.collection = db.collection(collectionName)
    return next()
  }
)
app.get('/', function (req, res) {
  res.send('please select a collection, '
    + 'e.g., /collections/messages')
})
app.get('/collections/:collectionName', function (req, res) {
  req.collection.find({}, {limit: 10, sort: [['_id', -1]]}).
    .toArray(function (e, results) {
      if (e) return next(e)
      res.send(results)
    }
  )
})

app.post('/collections/:collectionName', function (req, res) {
  req.collection.insert(req.body, {}, function (e, results) {
    if (e) return next(e)
    res.send(results)
  })
})
app.get('/collections/:collectionName/:id', function (req, res) {
  req.collection.findOne({_id: req.collection.id(req.params.id)},
    function (e, result) {
      if (e) return next(e)
      res.send(result)
    }
  )
})
app.put('/collections/:collectionName/:id', function (req, res) {
  req.collection.update({_id: req.collection.id(req.params.id)},
    {$set: req.body},
    {safe: true, multi: false},
```

8

```
      function (e, result) {
        if (e) return next(e)
        res.send((result === 1) ? {msg: 'success'} : {msg: 'error'})
      }
    )
  })
  app.del('/collections/:collectionName/:id', function (req, res) {
    req.collection.remove({_id: req.collection.id(req.params.id)},
      function (e, result) {
        if (e) return next(e)
        res.send((result === 1) ? {msg: 'success'} : {msg: 'error'})
      }
    )
  })

  app.listen(3000)
```

退出你的编辑器，在终端里运行：

```
$ node express.js
```

在另一个窗口（不要关闭之前的那个）运行：

```
$ mocha express.test.js
```

如果你不喜欢 Mocha 或者 BDD，CURL 也是可以使用的。

比如，使用 CURL 发送一个 `POST` 请求：

```
$ curl -d "" http://localhost:3000
```

`GET` 请求在浏览器里同样可用，比如 http://localhost:3000/test。

在这个例子里，我们的测试文件比应用的代码还长，看上去测试驱动应该被抛弃，但是请相信我，测试驱动的好习惯会在开发复杂应用时帮你节约很多时间。

8.7.4 总结

当你需要使用几行代码创建一个简单的 REST API 服务器的时候，Express.js 和 Mongoskin 都是非常棒的库。稍后，如果你想增强这个服务器，它们还提供了配置和组织代码的方式。

像 MongoDB 这样的 NoSQL 数据库对自由的 REST API 是有益的，我们不需要定义 schemas，并且可以把任何数据给它们保存。

完整的测试和应用文件，请浏览：https://gist.github.com/azat-co/6075685。

如果你想学习更多关于 Express.js 和其他 JavaScript 库以及相关知识，可以看看这个系列：Intro to Express.js tutorials[1]。

---

① http://webapplog.com/tag/intro-to-express-js/

注意

在这个例子里，我使用了不加分号的代码格式。在 JavaScript 里分号是可选的[1]，以下两种情况除外：一种是在 `for` 循环里，另一种是在以括号开头的表达式或者语句里（比如：立即调用的函数表达式[2]）。

8.8 Node.js MVC：Express.js + Derby Hello World 教程

8.8.1 Node MVC框架

Express.js[3]是一个流行 node 框架，它使用中间件来增强应用的功能。Derby[4]是一个 MVC[5]框架，使用时以 Express[6]为中间件。Derby 还支持 Racer[7]数据同步引擎，类似于 Handlebars[8]的模板引擎，以及很多其他功能[9]。

8.8.2 Derby安装

让我们在不使用支架的情况下设置一个 Derby 应用。通常项目生成器会使学习一个全新的框架感觉困惑。这是一个完全简单的 Hello World 应用教程，但是它也拥有 Derby 的骨架和使用 websockets 的实时模板。

当然，我们需要 Node.js[10]和 NPM[11]，可以从 nodejs.org[12]获得。全局安装 Derby，可以执行：

```
$ npm install -g derby
```

检查 Derby 是否安装：

```
$ derby -V
```

写这本书时是 2013 年 4 月，我使用的版本是 0.3.15。我们应该为创建第一个应用做好准备了！

① http://blog.izs.me/post/2353458699/an-open-letter-to-JavaScript-leaders-regarding
② http://en.wikipedia.org/wiki/Immediately-invoked_function_expression
③ http://expressjs.com
④ http://derbyjs.com
⑤ http://en.wikipedia.org/wiki/Model%E2%80%93vieco%E2%80%93controller
⑥ http://expressjs.com
⑦ https://github.com/codeparty/racer
⑧ https://github.com/wycats/handlebars.js/
⑨ http://derbyjs.com/#features
⑩ http://nodejs.org
⑪ http://nodejs.org
⑫ http://nodejs.org

8.8.3　文件结构

这个项目的目录结构如下：

```
project/
  -package.json
  -index.js
  -derby-app.js
  views/
    derby-app.html
  styles/
    derby-app.less
```

8.8.4　依赖

让我们来把一些依赖和基本信息放到package.json文件里：

```
{
  "name": "DerbyTutorial",
  "description": "",
  "version": "0.0.0",
  "main": "./server.js",
  "dependencies": {
    "derby": "*",
    "express": "3.x"
  },
  "private": true
}
```

现在，可以运行`npm install`，它会把所有的依赖下载到node_modules目录。

8.8.5　视图

视图必须放在views目录，并且它们要么在与你的derby应用js文件名同名的目录里的index.html文件中，即views/derby-app/index.html，要么在与derby应用js同名的文件里，如derby-app.html。

在这个Hello World应用中，我们使用`<Body:>`模板和`{message}`变量。Derby使用和handlebars类似语法的模板mustach[1]来重新渲染绑定。index.html是这样的：

```
<Body:>
  <input value="{message}"><h1>{message}</h1>
```

Stylus/LESS文件也同样如此；在我们的例子里，index.css只有一行：

① http://mustache.github.io/

```
h1 {
  color: blue;
}
```

更多精彩的 CSS 预处理器，请查看文档 Stylus[1]和 LESS[2]。

8.8.6 主服务器

index.js 是我们的主服务器文件，我们用 `require()` 函数引入依赖来开始这个文件：

```
var http = require('http'),
  express = require('express'),
  derby = require('derby'),
  derbyApp = require('./derby-app');
```

最后一行是我们的 Derby 应用 derby-app.js。

现在我们创建 Express.js 应用（3.*x* 与 2.*x* 版本有重大区别）和一个 HTTP 服务器：

```
var expressApp = new express(),
  server = http.createServer(expressApp);
```

Derby[3]和 Racer[4]数据同步库，我们这样创建：

```
var store = derby.createStore({
  listen: server
});
```

从后端获取数据到前端，我们需要实例化模型对象：

```
var model = store.createModel();
```

最重要的事情是需要把模型和路由作为中间件传给 Express.js 应用。我们需要指明 public 目录，以便 socket.io 正常工作：

```
expressApp.
  use(store.modelMiddleware()).
  use(express.static(__dirname + '/public')).
  use(derbyApp.router()).
  use(expressApp.router);
```

现在可以在端口 3001（或任意其他端口）打开服务器：

```
server.listen(3001, function () {
  model.set('message', 'Hello World!');
});
```

① http://learnboost.github.io/stylus/

② http://lesscss.org/

③ http://derbyjs.com

④ https://github.com/codeparty/racer

完整的 index.js 源码：

```
var http = require('http'),
  express = require('express'),
  derby = require('derby'),
  derbyApp = require('./derby-app');

var expressApp = new express(),
  server = http.createServer(expressApp);

var store = derby.createStore({
  listen: server
});

var model = store.createModel();

expressApp.
  use(store.modelMiddleware()).
  use(express.static(__dirname + '/public')).
  use(derbyApp.router()).
  use(expressApp.router);

server.listen(3001, function () {
  model.set('message', 'Hello World!');
});
```

8.8.7 Derby应用

最终，这个 Derby 应用，同时包含了前端代码和后端代码。前端代码在 app.ready() 回调里。启动它，我们来加载并创建一个应用。Derby 使用了一种不常见的初始化（不太像以前的 module.exports = app）：

```
var derby = require('derby'),
  app = derby.createApp(module);
```

为了使 socket.io 工作，我们需要订阅模型的属性，换言之，绑定数据与视图。我们可以在根路由里设置，下面是我们如何定义的：

```
app.get('/', function (page, model, params) {
  model.subscribe('message', function () {
    page.render();
  });
});
```

完整的 derby-app.js 文件：

```
var derby = require('derby'),
  app = derby.createApp(module);

app.get('/', function (page, model, params) {
  model.subscribe('message', function () {
    page.render();
  });
});
```

8.8.8　运行Hello World应用

现在万事俱备，可以启动我们的服务器了。运行 `node .`或者 `node index.js`，然后用浏览器打开 localhost:3001[①]。你会看到类似的东西：

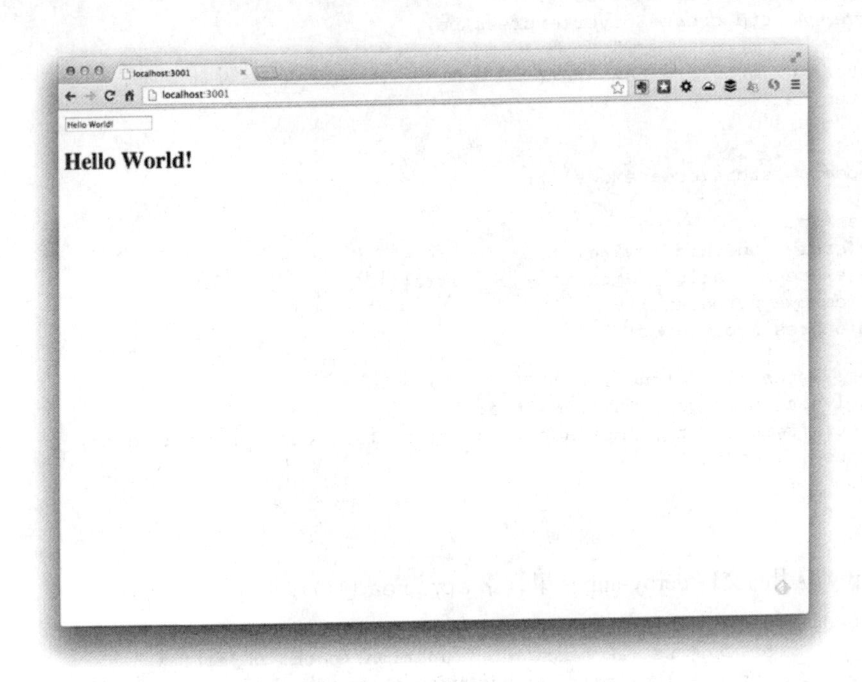

Derby + Express.js Hello World 应用

8.8.9　递值给后端

当然，不可变的数据不是很好，我们可以稍微修改一下应用让前端和后端进行一点点的交流。

在服务器文件 index.js 里，添加 `store.afterDb` 监听 `message` 属性的 `set` 事件：

```
server.listen(3001, function () {
  model.set('message', 'Hello World!');
  store.afterDb('set', 'message', function (txn, doc, prevDoc, done) {
    console.log(txn)
    done();
  });
});
```

修改后完整的 index.js 的代码：

① http://localhost:3001

```
var http = require('http'),
  express = require('express'),
  derby = require('derby'),
  derbyApp = require('./derby-app');

var expressApp = new express(),
  server = http.createServer(expressApp);

var store = derby.createStore({
  listen: server
});

var model = store.createModel();

expressApp.
  use(store.modelMiddleware()).
  use(express.static(__dirname + '/public')).
  use(derbyApp.router()).
  use(expressApp.router);

server.listen(3001, function () {
  model.set('message', 'Hello World!');
  store.afterDb('set', 'message', function (txn, doc, prevDoc, done) {
    console.log(txn)
    done();
  });
});
```

在Derby应用的文件derby-app.js里，给`app.ready()`添加`model.on()`：

```
app.ready(function (model) {
          model.on('set', 'message', function (path, object) {
            console.log('message has been changed: ' + object);
        })
});
```

修改后的完整derby-app.js源码：

```
var derby = require('derby'),
  app = derby.createApp(module);

app.get('/', function (page, model, params) {
  model.subscribe('message', function () {
    page.render();
  })
});

app.ready(function (model) {
  model.on('set', 'message', function (path, object) {
    console.log('message has been changed: ' + object);
  })
});
```

现在我们可以同时在终端和浏览器的开发者工具里看到日志。最终的结果在浏览器里是这样的：

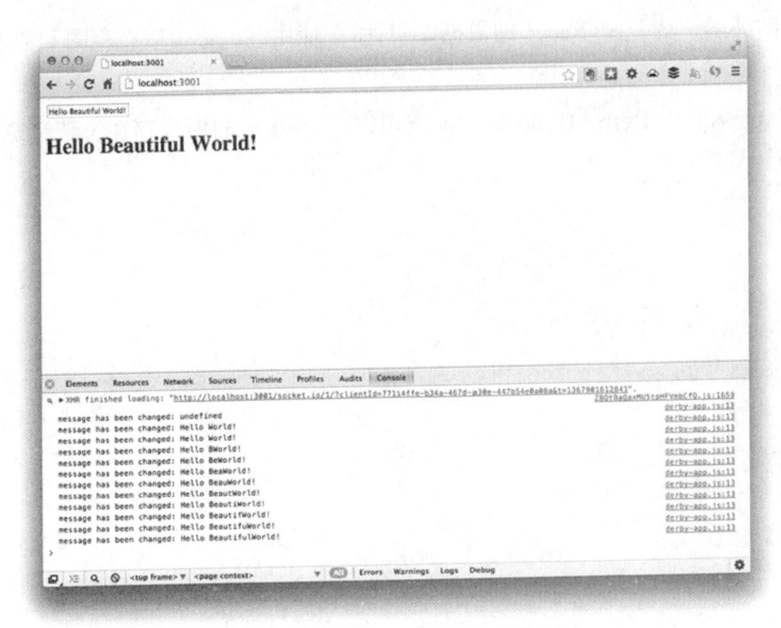

Hello World 应用：浏览器控制台日志

在终端里看上去如下：

```
3. node . (node)
[ 6,
  '77114ffe-b34a-467d-a30e-447b54e0a08a.18',
  'set',
  [ 'message', 'Hello BeautWorld!' ] ]
[ 7,
  '77114ffe-b34a-467d-a30e-447b54e0a08a.19',
  'set',
  [ 'message', 'Hello BeautiWorld!' ] ]
[ 8,
  '77114ffe-b34a-467d-a30e-447b54e0a08a.20',
  'set',
  [ 'message', 'Hello BeautifWorld!' ] ]
[ 9,
  '77114ffe-b34a-467d-a30e-447b54e0a08a.21',
  'set',
  [ 'message', 'Hello BeautifuWorld!' ] ]
[ 10,
  '77114ffe-b34a-467d-a30e-447b54e0a08a.22',
  'set',
  [ 'message', 'Hello BeautifulWorld!' ] ]
[ 11,
  '77114ffe-b34a-467d-a30e-447b54e0a08a.23',
  'set',
  [ 'message', 'Hello Beautiful World!' ] ]
```

Hello World 应用：终端控制台日志

更多神奇的展示，请查阅 Racer 的数据库属性[1]。使用它，你可以在视图和数据库间进行自动同步。

完整的 Express.js + Derby Hello World 应用以一个 gist 的形式存在，请浏览：gist.github.com/azat-co/5530311[2]。

① http://derbyjs.com/#persistence

② https://gist.github.com/azat-co/5530311

总结与推荐阅读

提要：总结本书，给出 JavaScript 博文、文章、电子书及其他资源的列表。

总结

希望你喜欢本书。知易行难，我们尝试展现多种多样的技术、框架以及现代 Web 敏捷开发中应用的技巧。本书涉及的主题包括：

- jQuery
- AJAX
- CSS 和 LESS
- JSON 和 BSON
- Twitter Bootstrap
- Node.js
- MongoDB
- Parse.com
- 敏捷方法
- Git
- Heroku、MongoHQ 和 Windows Azure
- REST API
- Backbone.js
- AMD 和 Require.js
- Express.js
- Monk
- Derby

如果你还想获取更有深度的知识或者参考资料，只需轻松点击一下相关的链接，或者用网上搜索一下即可。

构建不同版本 Chat 应用时包含的实际应用：

- jQuery + Parse.com JS REST API
- Backbone 和 Parse.com JS SDK
- Backbone 和 Node.js
- Backbone 和 Node.js + MongoDB

Chat 应用具备一般 Web 应用或者移动应用的各种基础功能：获取数据、展示数据、提交新数据。其他的例子包括：

- jQuery + Twitter RESP API “Tweet Analyzer”
- Parse.com “Save John”
- Node.js “Hello World”
- MongoDB “Print Collections”
- Derby + Express “Hello World”
- Backbone.js “Hello World”
- Backbone.js “Apple Database
- Monk + Expres.js “REST API Server”

如果你有任何反馈、评论、建议，或者发现拼写错误、程序错误、错误数据等，请提交一个 GitHub issue：https://github.com/azat-co/rpjs/issues。

其他联系方式：@azat_co[1]、http://webapplog.com、http://azat.co。

如果你喜欢 Node.js 并且想查看更多使用 Express.js 构建产品的内容，可以阅读我的新书 *Express.js Guide: The Most Popular Node.js Framework Manual*[2]。

推荐阅读

下面是一些资源、教程、图书和博客的列表，推荐大家阅读。

JavaScript资源和免费电子书

- *Oh My JS*[3]：最好的 JavaScript 文章精选集
- *JavaScript For Cats*[4]： 给新手程序员的介绍
- Eloquent JavaScript[5]：关于编程的时新介绍

---

① http://twitter.com/azat_co
② http://expressjsguide.com
③ https://leanpub.com/ohmyjs/read
④ http://jsforcats.com/
⑤ http://eloquentjavascript.net/

- Superhero.js[1]：JS 资源的全面整合
- JavaScript Guide[2]：Mozilla 开发者中心的 JavaScript 指南
- JavaScript Reference[3]：Mozilla 开发者中心的 JavaScript 参考
- Why Use Closure[4]：基于事件编程中闭包的实际应用
- Prototypal Inheritance[5]：关于对象继承与局部变量
- Control Flow in Node[6]：流程控制里并行与串行的对比
- Truthy and Falsey Values[7]
- How to Write Asynchronous Code[8]：怎么写异步代码
- *Smooth CoffeeScript*[9]：免费的交互式 HTML5 书籍，快速参考集合及其他一些好东西
- *Developing Backbone.js Applications*[10]：Addy Osmani 的早期免费版（O'Reilly）
- Step by Step from jQuery to Backbone[11]：从 jQuery 到 Backbone 的学习指南
- Open Web Platform Daily Digest[12]：JS 每日文摘
- DISTILLED HYPE[13]：JS 博客/时事

Javascript书籍

- *JavaScript：The Good Parts*[14]
- *JavaScript：The Definitive Guide*[15]
- *Secrets of the JavaScript Ninja*[16]
- *Pro JavaScript Techniques*[17]

---

① http://superherojs.com/
② https://developer.mozilla.org/en-US/docs/JavaScript/Guide
③ https://developer.mozilla.org/en-US/docs/JavaScript/Reference
④ http://howtonode.org/why-use-closure
⑤ http://howtonode.org/prototypical-inheritance
⑥ http://howtonode.org/control-flow
⑦ http://docs.nodejitsu.com/articles/javascript-conventions/what-are-truthy-and-falsy-values
⑧ http://docs.nodejitsu.com/articles/getting-started/control-flow/how-to-write-asynchronous-code
⑨ http://autotelicum.github.com/Smooth-CoffeeScript/
⑩ http://addyosmani.github.com/backbone-fundamentals/
⑪ https://github.com/kjbekkelund/writings/blob/master/published/understanding-backbone.md
⑫ http://daily.w3viewer.com/
⑬ http://distilledhype.com/
⑭ http://shop.oreilly.com/product/9780596517748.do
⑮ http://www.amazon.com/dp/0596101996/?tag=stackoverfl08-20
⑯ http://www.manning.com/resig/
⑰ http://www.amazon.com/dp/1590597273/?tag=stackoverfl08-20

Node.js资源和免费电子书

- Felix's Node.js Beginners Guide①
- Felix's Node.js Style Guide②
- Felix's Node.js Convincing the boss guide③
- Introduction to NPM④
- NPM Cheatsheet⑤
- Interactive Package.json Cheatsheet⑥
- Official Node.js Documentation⑦
- Node Guide⑧
- Node Tuts⑨
- *What Is Node?*⑩：Kindle 免费电子书
- *Mastering Node.js*⑪：node 开源电子书
- *Mixu's Node book*⑫：如何使用 Node.js
- *Learn Node.js Completely and with Confidence*⑬：两周自学 JavaScript 指南
- How to Node⑭：Node.js 编程之禅

Node.js书籍

- *The Node Beginner Book*⑮
- *Hands-on Node.js*⑯
- *Backbone Tutorials*⑰

① http://nodeguide.com/beginner.html
② http://nodeguide.com/style.html
③ http://nodeguide.com/convincing_the_boss.html
④ http://howtonode.org/introduction-to-npm
⑤ http://blog.nodejitsu.com/npm-cheatsheet
⑥ http://package.json.nodejitsu.com/
⑦ http://nodejs.org/api/
⑧ http://nodeguide.com/
⑨ http://nodetuts.com/
⑩ http://www.amazon.com/What-Is-Node-ebook/dp/B005ISQ7JC
⑪ http://visionmedia.github.com/masteringnode/
⑫ http://book.mixu.net/
⑬ http://javascriptissexy.com/learn-node-js-completely-and-with-confidence/
⑭ http://howtonode.org/
⑮ https://leanpub.com/nodebeginner
⑯ https://leanpub.com/hands-on-nodejs
⑰ https://leanpub.com/backbonetutorials

- ❑ *Smashing Node.js*[①]
- ❑ *The Node Beginner Book*[②]
- ❑ *Hands-on Node.js*[③]
- ❑ *Node: Up and Running*[④]
- ❑ *Node.js in Action*[⑤]
- ❑ *Node: Up and Running*[⑥]：使用 JavaScript 构建可伸缩的服务器端代码
- ❑ *Node Web Development*[⑦]：关于 Node 的实用介绍
- ❑ *Node Cookbook*[⑧]

在线互动课堂和教程

- ❑ Cody Academy[⑨]： 交互式编程教学
- ❑ Programr[⑩]
- ❑ LearnStreet[⑪]
- ❑ Treehouse[⑫]
- ❑ lynda.com[⑬]：软件、创意、商业课程
- ❑ Udacity[⑭]：大量在线公开课程
- ❑ Coursera[⑮]

创业的书和博客

- ❑ 《黑客与画家》
- ❑ 《精益创业》

---

① http://www.amazon.com/Smashing-Node-js-JavaScript-Everywhere-Magazine/dp/1119962595/
② http://www.nodebeginner.org/
③ http://nodetuts.com/handson-nodejs-book.html
④ http://shop.oreilly.com/product/0636920015956.do
⑤ http://www.manning.com/cantelon/
⑥ http://www.amazon.com/Node-Running-Scalable-Server-Side-JavaScript/dp/1449398588
⑦ http://www.amazon.com/Node-Web-Development-David-Herron/dp/184951514X
⑧ http://www.amazon.com/Node-Cookbook-David-Mark-Clements/dp/1849517185/
⑨ http://www.codecademy.com/
⑩ http://www.programr.com/
⑪ http://www.learnstreet.com/
⑫ http://teamtreehouse.com/
⑬ http://www.lynda.com/
⑭ https://www.udacity.com/
⑮ https://www.coursera.org/

- 《创业者手册》
- *The Entrepreneur's Guide to Customer Development: A cheat sheet to The Four Steps to the Epiphany*[1]
- Venture Hacks[2]
- WebAppLog[3]

① http://www.amazon.com/The-Entrepreneurs-Guide-Customer-Development/dp/0982743602/
② http://venturehacks.com/
③ http://webapplog.com

图灵最新重点图书

- 程序员爸爸的第一本亲子互动编程书
- 腾讯效果广告平台部商务研发中心总监陈俊
  全国青少年信息学奥林匹克竞赛金牌教练曹文
  联袂推荐
- 内容经过教育专家的评审，经过孩子的亲身检验，并得到了家长的认可

父与子的编程之旅
书号：978-7-115-36717-4
作者：Warren Sande Carter Sande
定价：69.00 元

Python 计算机视觉编程
书号：978-7-115-35232-3
作者：Jan Erik Solem
定价：69.00 元

Python 开发实战
书号：978-7-115-32089-6
作者：BePROUD 股份有限公司
定价：79.00 元

Python 基础教程（第 2 版·修订版）
书号：978-7-115-35352-8
作者：Magnus Lie Hetland
定价：79.00 元

程序员必读之软件架构
书号：978-7-115-37107-2
作者：Simon Brown
定价：49.00 元

AngularJS 权威教程
书号：978-7-115-36647-4
作者：Ari Lerner
定价：99.00 元

数据结构与算法 JavaScript 描述
书号：978-7-115-36339-8
作者：Michael McMillan
定价：49.00 元

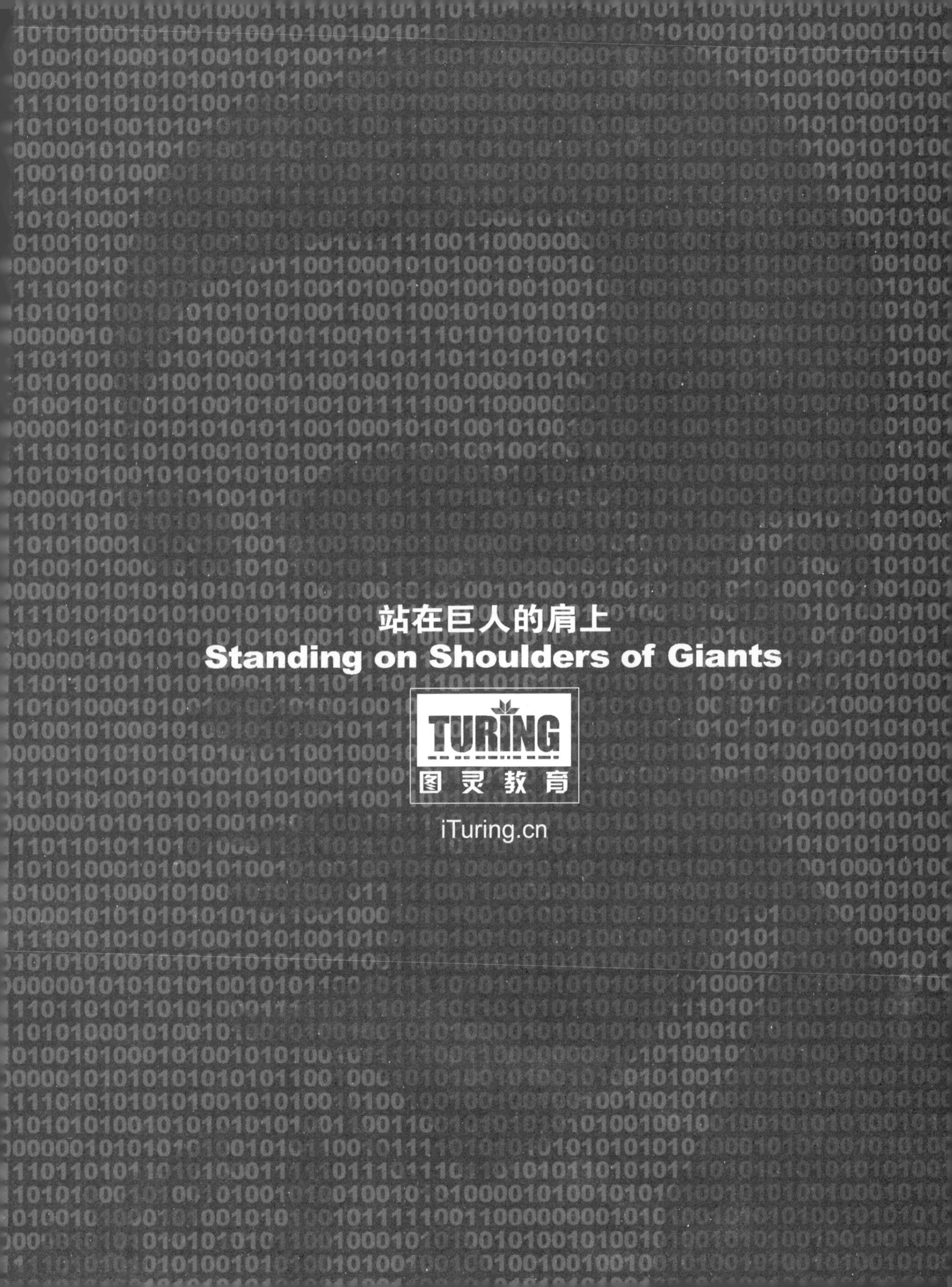

站在巨人的肩上
Standing on Shoulders of Giants
TURING
图灵教育
iTuring.cn

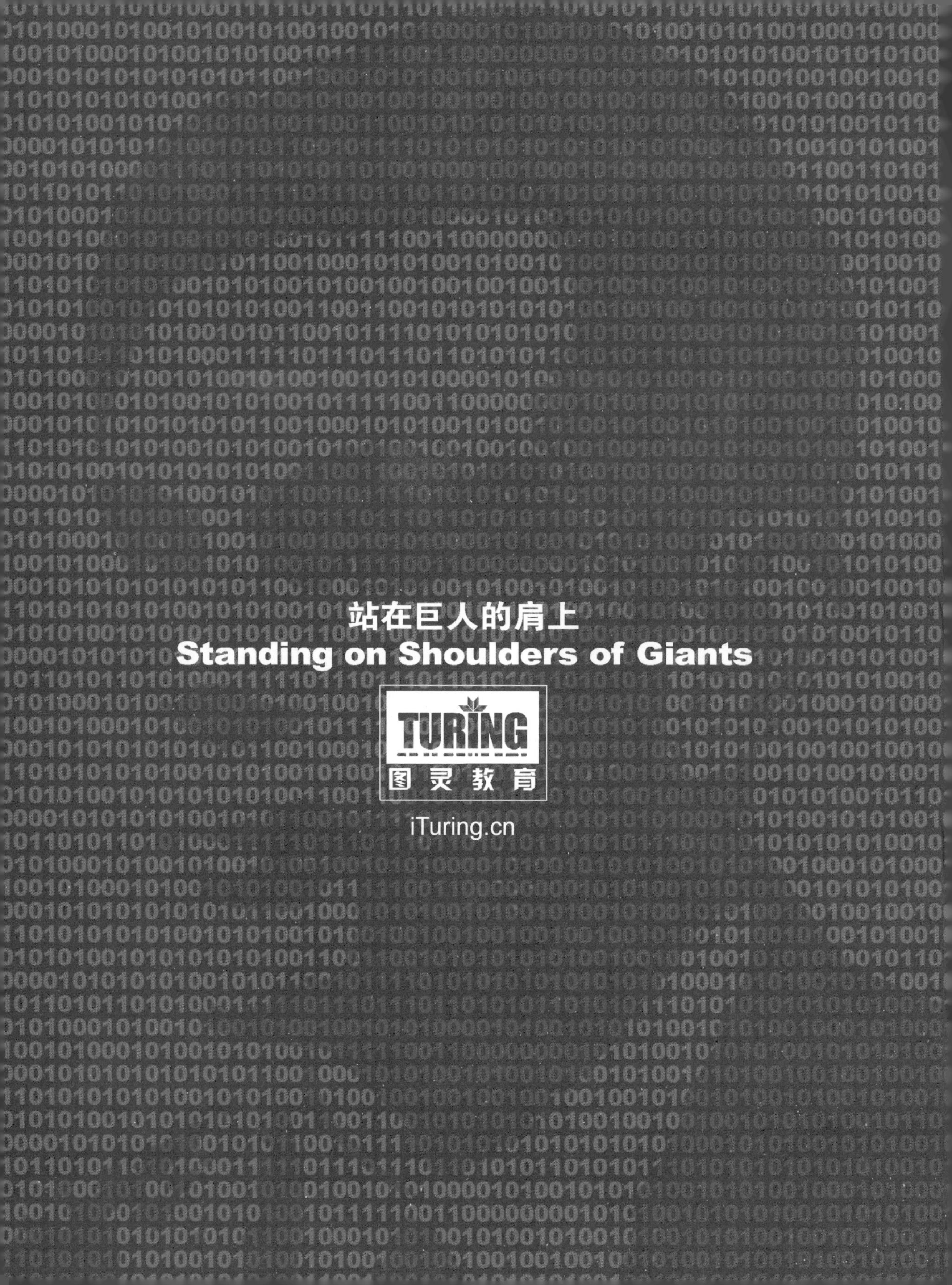
站在巨人的肩上
Standing on Shoulders of Giants
TURING
图灵教育
iTuring.cn